MEASURING AND SUSTAINING THE NEW ECONOMY

ENHANCING PRODUCTIVITY GROWTH IN THE INFORMATION AGE

Committee on Measuring and Sustaining the New Economy

Board on Science, Technology, and Economic Policy

Policy and Global Affairs

Dale W. Jorgenson and Charles W. Wessner, *Editors*

NATIONAL RESEARCH COUNCIL
OF THE NATIONAL ACADEMIES

THE NATIONAL ACADEMIES PRESS
Washington, D.C.
www.nap.edu

THE NATIONAL ACADEMIES PRESS 500 Fifth Street, N.W. Washington, DC 20001

NOTICE: The project that is the subject of this report was approved by the Governing Board of the National Research Council, whose members are drawn from the councils of the National Academy of Sciences, the National Academy of Engineering, and the Institute of Medicine. The members of the committee responsible for the report were chosen for their special competences and with regard for appropriate balance.

This study was supported by: Contract/Grant No. CMRC-50SBNB9C1080 between the National Academy of Sciences and the U.S. Department of Commerce; Contract/Grant No. NASW-99037, Task Order 103, between the National Academy of Sciences and the National Aeronautics and Space Administration; Contract/Grant No. CMRC-SB134105C0038 between the National Academy of Sciences and the U.S. Department of Commerce; Contract/Grant No. OFED-13416 between the National Academy of Sciences and Sandia National Laboratories; Contract/Grant No. N00014-00-G-0230, DO #23, between the National Academy of Sciences and the Department of the Navy; Contract/Grant No. NSF-EIA-0119063 between the National Academy of Sciences and the National Science Foundation; and Contract/Grant No. DOE-DE-FG02-01ER30315 between the National Academy of Sciences and the U.S. Department of Energy. Additional support was provided by Intel Corporation. Any opinions, findings, conclusions, or recommendations expressed in this publication are those of the author(s) and do not necessarily reflect the views of the organizations or agencies that provided support for the project.

International Standard Book Number-10: 0-309-10220-0
International Standard Book Number-13: 978-0-309-10220-9

Limited copies are available from Board on Science, Technology, and Economic Policy, National Research Council, 500 Fifth Street, N.W., W547, Washington, DC 20001; 202-334-2200.

Additional copies of this report are available from the National Academies Press, 500 Fifth Street, N.W., Lockbox 285, Washington, DC 20055; (800) 624-6242 or (202) 334-3313 (in the Washington metropolitan area); Internet, http://www.nap.edu

Printed in the United States of America

THE NATIONAL ACADEMIES

Advisers to the Nation on Science, Engineering, and Medicine

The **National Academy of Sciences** is a private, nonprofit, self-perpetuating society of distinguished scholars engaged in scientific and engineering research, dedicated to the furtherance of science and technology and to their use for the general welfare. Upon the authority of the charter granted to it by the Congress in 1863, the Academy has a mandate that requires it to advise the federal government on scientific and technical matters. Dr. Ralph J. Cicerone is president of the National Academy of Sciences.

The **National Academy of Engineering** was established in 1964, under the charter of the National Academy of Sciences, as a parallel organization of outstanding engineers. It is autonomous in its administration and in the selection of its members, sharing with the National Academy of Sciences the responsibility for advising the federal government. The National Academy of Engineering also sponsors engineering programs aimed at meeting national needs, encourages education and research, and recognizes the superior achievements of engineers. Dr. Wm. A. Wulf is president of the National Academy of Engineering.

The **Institute of Medicine** was established in 1970 by the National Academy of Sciences to secure the services of eminent members of appropriate professions in the examination of policy matters pertaining to the health of thc public. The Institute acts under the responsibility given to the National Academy of Sciences by its congressional charter to be an adviser to the federal government and, upon its own initiative, to identify issues of medical care, research, and education. Dr. Harvey V. Fineberg is president of the Institute of Medicine.

The **National Research Council** was organized by the National Academy of Sciences in 1916 to associate the broad community of science and technology with the Academy's purposes of furthering knowledge and advising the federal government. Functioning in accordance with general policies determined by the Academy, the Council has become the principal operating agency of both the National Academy of Sciences and the National Academy of Engineering in providing services to the government, the public, and the scientific and engineering communities. The Council is administered jointly by both Academies and the Institute of Medicine. Dr. Ralph J. Cicerone and Dr. Wm. A. Wulf are chair and vice chair, respectively, of the National Research Council.

www.national-academies.org

Committee on Measuring and Sustaining the New Economy*

Dale Jorgenson, *Chair*
Samuel W. Morris University Professor
Harvard University

William J. Spencer, *Vice Chair*
Chairman Emeritus, *retired*
SEMATECH

M. Kathy Behrens
Managing Director of Medical Technology
Robertson Stephens Investment Management

Kenneth Flamm
Dean Rusk Chair in International Affairs
Lyndon B. Johnson School of Public Affairs
University of Texas at Austin

Bronwyn Hall
Professor of Economics
University of California at Berkeley

James Heckman
Henry Schultz Distinguished Service Professor of Economics
University of Chicago

Richard Levin
President
Yale University

David T. Morgenthaler
Founding Partner
Morgenthaler Ventures

Mark B. Myers
Visiting Executive Professor of Management
The Wharton School
University of Pennsylvania

Roger Noll
Morris M. Doyle Centennial Professor of Economics
Director, Public Policy Program
Stanford University

Edward E. Penhoet
President
Gordon and Betty Moore Foundation

William Raduchel
Chairman and CEO
The Ruckus Network

Alan Wm. Wolff
Managing Partner
Dewey Ballantine

*As of June 2006.

Project Staff*

Charles W. Wessner
Study Director

McAlister T. Clabaugh
Program Associate

David E. Dierksheide
Program Officer

Paul J. Fowler
Senior Research Associate

Jeffrey McCullough
Program Associate

Sujai J. Shivakumar
Program Officer

*As of June 2006.

For the National Research Council (NRC), this project was overseen by the Board on Science, Technology, and Economic Policy (STEP), a standing board of the NRC established by the National Academies of Sciences and Engineering and the Institute of Medicine in 1991. The mandate of the STEP Board is to integrate understanding of scientific, technological, and economic elements in the formulation of national policies to promote the economic well-being of the United States. A distinctive characteristic of STEP's approach is its frequent interactions with public- and private-sector decision makers. STEP bridges the disciplines of business management, engineering, economics, and the social sciences to bring diverse expertise to bear on pressing public policy questions. The members of the STEP Board* and the NRC staff are listed below:

Dale Jorgenson, *Chair*
Samuel W. Morris University Professor
Harvard University

Timothy Bresnahan
Landau Professor in Technology and the Economy
Stanford University

Lewis Coleman
President
DreamWorks Animation

Kenneth Flamm
Dean Rusk Chair in International Affairs
Lyndon B. Johnson School of Public Affairs
University of Texas at Austin

Mary L. Good
Donaghey University Professor
Dean, Donaghey College of Information Science and Systems Engineering
University of Arkansas at Little Rock

Amo Houghton
Member of Congress (ret.)

David T. Morgenthaler
Founding Partner
Morgenthaler Ventures

Joseph Newhouse
John D. MacArthur Professor of Health Policy and Management
Harvard University

Edward E. Penhoet
President
Gordon and Betty Moore Foundation

Arati Prabhakar
General Partner
U.S. Venture Partners

William J. Raduchel
Chairman and CEO
The Ruckus Network

Jack Schuler
Chairman
Ventana Medical Systems

Suzanne Scotchmer
Professor of Economics and Public Policy
University of California at Berkeley

*As of June 2006.

STEP Staff*

Stephen A. Merrill
Executive Director

McAlister T. Clabaugh
Program Associate

David E. Dierksheide
Program Officer

Paul Fowler
Senior Research Associate

Charles W. Wessner
Program Director

Jeffrey McCullough
Program Associate

Ben Roberts
Christine Mirzayan Science
and Technology Policy Fellow

Sujai J. Shivakumar
Program Officer

*As of June 2006.

Contents

Preface

The New Economy refers to technological and structural changes in the U.S. economy as individuals capitalize on new technologies, new opportunities, and national investments in computing, information, and communications technologies. Ongoing rapid declines in the prices of computers and semiconductors as well as apparent similar declines in the prices of software and communications equipment have led to diverse new information technology (IT)-enabled capabilities and the widespread adoption of information technologies. These investments have significantly improved the nation's productivity, raising the trajectory of economic growth since the mid-1990s.[1] This gain appears to be robust, having survived the dot-com crash, the short recession of 2001, and the tragedy of 9/11. Since the end of the previous recession of 2001, productivity growth had been running at about two-tenths of a percentage point higher than in any recovery of the post-World War II period.[2]

A structural change most associated with the New Economy today is the transformation of the Internet from a communication media to a platform for service delivery.[3] This has led to the remarkable growth of the U.S. service

[1]Dale W. Jorgenson and Kevin J. Stiroh, "Raising the Speed Limit: Economic Growth in the Information Age," in National Research Council, *Measuring and Sustaining the New Economy*, Dale W. Jorgenson and Charles W. Wessner, eds., Washington, D.C.: The National Academies Press, 2002, Appendix A.

[2]Dale W. Jorgenson, Mun S. Ho, and Kevin J. Stiroh, "Will the U.S. Productivity Resurgence Continue?" *Federal Reserve Bank of New York Current Issues in Economics and Finance,* 10(13), 2004.

[3]This transformation is sometimes referred to as "Web 2.0." For a description of this new version of the Web, see Tim O'Reilly, "What Is Web 2.0—Design Patterns and Business Models

economy, as companies like Google and eBay increasingly exploit information services in new ways.

THE CONTEXT OF THIS REPORT: THE COMMITTEE'S TASK

In order to sustain the benefits of higher productivity and economic growth, policy makers need to improve their understanding of the operation of this new American economy. Unfortunately, the empirical record of this change is incomplete, with much remaining to be done before definitive quantitative assessments can be made about the role that these new technological assets play in the U.S. economy.

To meet this need, the Board on Science, Technology, and Economic Policy (STEP) appointed a committee to convene a series of conferences designed to identify and address the policy issues associated with the measurement, development, and growth characteristic of the "New Economy." Focusing primarily on the Information Technology sector as the major driver of productivity growth from the 1990s, the study has examined key sectors, or building blocks, that underpin this new and more productive U.S. economy. These sectors include semiconductors (one of the principal drivers of productivity growth), computers and their various components (another driver of productivity), software in its various forms (pervasive throughout the economy), and the contributions of dramatically improved capacity and lower cost of telecommunication services and data transmission.

The study took up a series of issues such as:

- measurement issues, such as data classification and collection requirements;
- the building block technologies of the "new economy," including their special characteristics and synergies across industries; and
- policy and regulatory issues.

CONFERENCES ON THE NEW ECONOMY

Following an initial conference that provided the impetus for this project, separate conferences on each of these sectors were convened over several years. Each identified major issues associated with the measurement and analysis of the current U.S. economy, the technologies underpinning its growth, and the government-industry collaborations and regulatory framework necessary to sustain its continued advance. The proceedings of each of these conferences have been captured in separate reports. These reports, together with commissioned

---

for the Next Generation of Software" September 30, 2005. Accessed at <*http://www.oreillynet.com/pub/a/oreilly/tim/news/2005/09/30/what-is-web-20.html*>.

papers, have been used to establish a basis for this final consensus report by the Committee.

The conferences included:

- **Measuring and Sustaining the New Economy.** Held on October 6, 2000, at the National Academies, Washington, D.C., this initial exploratory conference described the nature and sources of growth in the "new" economy and set out the broad challenges faced in measuring and sustaining the growth in productivity that characterizes it. The conference provided the initial impetus for this project and, with subsequent approval and funding, led to a series of workshops dealing with the specific sectors most closely linked to emergence and sustainability of the positive trends that now distinguish the U.S. economy. An initial report captured the deliberations of this conference and attracted the interest of Washington policy makers.[4]
- **Productivity and Cyclicality in the Semiconductor Industry.** Held at Harvard University on September 24, 2001, soon after the 11 September terror attacks, this conference looked at the trends, implications, and policy questions that arise from an understanding of the Moore's Law phenomenon in semiconductors. It explored how the cyclicality found in the semiconductor industry might be modeled. It also highlighted a variety of policy initiatives needed to help sustain a vibrant semiconductor industry in the United States.[5]
- **Deconstructing the Computer.** This conference, held on February 28, 2003, brought together leading figures from the different industries that develop and manufacture computer components (such as printers, memories, and monitors) to examine the extent of Moore's Law phenomenon in their industry and to explore how best to measure computer performance and how to sustain the benefits to the economy arising from a Moore's Law for computer components.[6]
- **Software, Growth, and the Future of the U.S. Economy.** This conference, held on February 20, 2004, examined the nature of software, reviewed how software has been measured in the national accounts, and discussed the challenges of capturing the value and vulnerabilities of software, given that the nature of software is itself rapidly evolving. The meeting also highlighted the globalization of the software industry, the recent "offshoring" phenomenon, and the policy challenges that these developments pose.[7]

---

[4]See National Research Council, *Measuring and Sustaining the New Economy*, Dale W. Jorgenson and Charles W. Wessner, eds., Washington, D.C.: National Academy Press, 2002.

[5]See National Research Council, *Productivity and Cyclicality in Semiconductors: Trends, Implications, and Questions,* Dale W. Jorgenson and Charles W. Wessner, eds., Washington, D.C.: The National Academies Press, 2004.

[6]See National Research Council, *Deconstructing the Computer,* Dale W. Jorgenson and Charles W. Wessner, eds., Washington, D.C.: The National Academies Press, 2005.

[7]See National Research Council, *Software, Growth, and the Future of the U.S. Economy,* Dale W. Jorgenson and Charles W. Wessner, eds., Washington, D.C.: The National Academies Press, 2006.

• **The Telecommunications Challenge: Changing Technologies and Evolving Policies.** Taking stock of the rapid convergence between telecommunications and information technologies, participants at this final conference in the series, held on November 15, 2004, examined the need to expand the reach of the nation's high-bandwidth broadband network. Discussed in this context was the need for an adaptive policy framework that encourages innovation in telecommunications, and the development of new business models for telephony as well as voice and video entertainment.[8]

The proceedings of each of these conferences have been published in separate volumes by The National Academies Press. Although the technologies of the industries considered at these conferences continue to evolve rapidly, the reports nonetheless capture conceptual issues of continued policy relevance to the industry leaders, academics, policy analysts, and others who participated in these workshops. Part III of this report summarizes key issues taken up at these five conferences. The knowledge and insights reflected in the remarks of industry and policy representatives capture tacit knowledge that is not always available through formal academic analysis. These insights, buttressed by the commissioned papers, provide a valuable review and, in some cases, analysis of selected themes. Both the workshop summaries and the papers contributed to this consensus report.

While this report reflects the wide scope of the Committee's deliberations, it does not attempt (nor could it hope) to address all aspects of the causes and effects of the modern information economy. For instance, this volume does not provide a comprehensive study of the developments within industries that use information technologies, such as the retailing or transportation industries. Nor does it provide a detailed discussion of the experience of information technology-supplying industries that are related to but outside the Internet, such as the supercomputing or cellular industries. Other areas not covered include the potential role of standards committees in fostering technical advance, possible changes to the Generally Accepted Accounting Principles to improve how R&D is captured, the role nanotechnologies could play in advancing information technology, and changes in spectrum policy to move the spectrum from low value to higher value uses.

ACKNOWLEDGMENTS

There is considerable interest in the policy community in developing a better understanding of the technological drivers and appropriate regulatory framework

[8]See National Research Council, *The Telecommunications Challenge: Changing Technologies and Evolving Policies,* Charles W. Wessner, ed., Washington, D.C.: The National Academies Press, 2006.

for the New Economy, as well as a better grasp of its operation. This interest is reflected in the support on the part of agencies that have played a role in the creation and development of the technologies and regulatory frameworks that underpin the New Economy. We are grateful for the participation and the contributions of the National Science Foundation, the National Institute of Standards and Technology, the Department of Energy, the National Aeronautics and Space Administration, the Office of Naval Research, and Sandia National Laboratories.

Among STEP staff, we especially wish to thank Dr. Sujai Shivakumar for his instrumental role in the creation of this report. His ability to synthesize the diverse perspectives into a coherent whole while capturing key themes was essential. We are also indebted to David Dierksheide for his role in preparing this report for publication.

NRC REVIEW

This report has been reviewed in draft form by individuals chosen for their diverse perspectives and technical expertise, in accordance with procedures approved by the National Academies' Report Review Committee. The purpose of this independent review is to provide candid and critical comments that will assist the institution in making its published report as sound as possible and to ensure that the report meets institutional standards for objectivity, evidence, and responsiveness to the study charge. The review comments and draft manuscript remain confidential to protect the integrity of the process.

We wish to thank the following individuals for their review of this report: Ana Aizcorbe, Bureau of Economic Analysis; Bruce Grimm, Bureau of Economic Analysis; Paul Horn, IBM Thomas Watson Research Center; Way Kuo, University of Tennessee; William Scherlis, Carnegie Mellon University; Kevin Stiroh, Federal Reserve Bank of New York; William Taylor, NERA Economic Consulting; and Larry Thompson, Ultratech Stepper, Inc.

Although the reviewers listed above have provided many constructive comments and suggestions, they were not asked to endorse the conclusions or recommendations, nor did they see the final draft of the report before its release. The review of this report was overseen by Charles Phelps, University of Rochester. Appointed by the National Academies, he was responsible for making certain that an independent examination of this report was carried out in accordance with institutional procedures and that all review comments were carefully considered. Responsibility for the final content of this report rests entirely with the authoring committee and the institution.

STRUCTURE

Part I of this report is an introduction to the features and challenges of the New Economy by Dale Jorgenson. Part II of this report provides a summary of the Committee's findings and recommendations for each of the sectors covered in the series of conferences. Finally, Part III summarizes the main themes from the proceedings of the five conferences listed above, drawing together the main policy challenges for the United States in sustaining the productivity growth and improved welfare associated with the New Economy. The overview of the findings and recommendations that follows this preface is designed to provide the harried reader an overview of the new economy story as a whole, while Part II provides interested individuals a greater focus on the individual high-technology sectors that contribute to the remarkable growth of the United States economy.

Dale W. Jorgenson

List of Acronyms

BEA	Bureau of Economic Analysis of the U.S. Department of Commerce
BLS	Bureau of Labor Statistics of the U.S. Department of Labor
CD	Compact Disc
CMOS	Complementary Metal-Oxide Semiconductor, a major class of integrated circuits
DRAM	Dynamic Random Access Memory
DSL	Digital Subscriber Line is a family of technologies that provide digital data transmission over the wires of a local telephone network.
DVD	Digital Video Disk
EUV	Extreme Ultraviolet Lithography technology, capable of creating nanometer-scale patterns for use in semiconductor manufacturing
FASB	The Financial Accounting Standards Board (FASB) develops Generally Accepted Accounting Principles in the United States.
FCC	Federal Communications Commission
H-1B	H-1B is a U.S. visa category that allows American companies and universities to employ foreign scientists, engineers, programmers, and other professionals in the United States.
IEEE	Institute of Electrical and Electronics Engineers
IMEC	The Interuniversity Microelectronics Centre is a microelectronics research facility on the outskirts of Leuven, Belgium.
IPTV	Internet Protocol Television describes a system where a digital television service is delivered to subscribing consumers using the Internet Protocol over a broadband connection.
ITRS	International Technology Roadmap for Semiconductors

LCD Liquid Crystal Display

LLU Local Loop Unbundling is the process of allowing telecommunications operators to use the twisted-pair telephone connections from the telephone exchange's central office to the customer premises.

NIPA National Income and Product Accounts use double entry accounting to report the monetary value and sources of output produced in a country and the distribution of incomes that production generates.

OECD Organisation for Economic Co-operation and Development

OLED Organic Light Emitting Diode

SEC Securities and Exchange Commission

TCP/IP The Transmission Control Protocol (TCP) and the Internet Protocol (IP) are sets of communications protocols that implement the protocol stack on which the Internet and most commercial networks run.

VAT Value-Added Tax

VOIP Voice over Internet Protocol is the routing of voice conversations over the Internet or through any other IP-based network.

WDM Wavelength-division multiplexing is a technology that multiplexes multiple optical carrier signals on a single optical fiber by using different wavelengths (colors) of laser light to carry different signals.

WiFi A technology for wireless local area networks. Designed to be used for mobile computing devices such as laptops, it is increasingly used for applications including Internet access, gaming, and basic connectivity of consumer electronics such as televisions and DVD players.

WiMAX Worldwide Interoperability for Microwave Access. WiMAX is a standards-based wireless technology that provides high-throughput broadband connections over long distances. WiMAX can be used for a number of applications, including "last mile" broadband connections, hotspots and cellular backhaul, and high-speed enterprise connectivity for business.

WTO World Trade Organization

Overview of the Findings and Recommendations

Faster, better, and cheaper semiconductors and computers as well as software and telecommunications equipment have led to their widespread adoption and networked use. These innovations are creating fundamental changes to the means and speed with which knowledge is created, services are delivered, and goods are manufactured and distributed around the world. Information and communications technologies have transformed how individuals and corporate entities everywhere consume, work, interact, and transact. From the U.S. perspective, these changes are improving productivity and are raising the long-term growth trajectory of the U.S. economy.[1] Sustaining this growth, in turn, requires new approaches to economic measurement and policy analysis.

The challenge discussed in this volume is threefold: (1) to understand the diverse sources of these growth-enhancing productivity gains; (2) to better measure the contributions of different elements of the "new economy" story—that is to say, semiconductors, computers, software, and telecommunications; and (3) to develop policies to (i) meet the needs of these growth-enhancing industries and thereby benefit from their positive effects on the rest of the economy, and (ii) enable the United States to remain an attractive location for these industries within an increasingly competitive global economy.

[1]This gain in the growth rate appears to be robust, having survived the dot-com crash, the short recession of 2001, and the tragedy of 9/11. Since the end of the previous recession in 2001, productivity growth has been running at about two-tenths of a percentage point higher than in any recovery of the post-World War II period. See Dale W. Jorgenson, Mun S. Ho, and Kevin J. Stiroh, "Will the U.S. Productivity Resurgence Continue?" *Federal Reserve Bank of New York Current Issues in Economics and Finance,* 10(13), 2004.

This new global economy poses new opportunities as well as new challenges to the United States' growth and competitiveness. To better understand these trends, and the conditions to sustain them, the National Academies' Committee on Measuring and Sustaining the New Economy held a series of conferences since 2001 covering the semiconductor, computer component, software, and telecommunications sectors. Each of these conferences brought together industry leaders, economists, national accountants, and leading policy analysts to identify the data challenges and policies needed to sustain the advantages of this new technological and economic paradigm. This concluding report presents the Committee's findings and recommendations on the steps required to better understand what is happening to the U.S. economy (through better measurement) and policy measures that are needed to measure and sustain the benefits of this "new economy" within each of these sectors. This overview draws together the common themes from these sector-specific findings and recommendations.

SUSTAINING THE NEW ECONOMY

Following the expectations set by Moore's Law, semiconductors have been a driver of price-performance improvements in information technology. Declines in cost for electronics functionality embedded in semiconductors are the basis of improvements in price-performance in computers and communications equipment, which in turn has been a major factor in increasing the productivity and long-term growth performance of the U.S. economy.[2] Parallel improvements in the capacity of communications equipment, described as Gilder's Law, suggest that the maximum transmission rate for telecoms is tripling every year. Combined with Moore's Law, which forecasts that computer power doubles every 18 months, Gilder's Law implies that communications power doubles every 6 months.

While not pretending to be deterministic, Moore's formulation has endured in part by setting expectations among participants in the semiconductor industry of the pace of innovation and introduction of new products to market.[3] In as far as each firm believes that its competitors will release the next technological version in an 18-month timeframe, each firm tends to accelerate the pace of its own

[2]Jack E. Triplett, "High-Tech Productivity and Hedonic Price Indexes," in Organisation for Economic Co-operation and Development, *Industry Productivity,* Paris: Organisation for Economic Co-operation and Development, 1996; Kenneth Flamm, "Technological Advance and Costs: Computers vs. Communications," in *Changing the Rules: Technological Change, International Competition, and Regulation in Communications,* Robert C. Crandall and Kenneth Flamm, eds., Washington, D.C.: The Brookings Institution, 1989; Ana Aizcorbe, Kenneth Flamm, and Anjum Khurshid, "The Role of Semiconductor Inputs in IT Hardware Price Declines," in *Hard to Measure Goods and Services: Essays in Honor of Zvi Griliches,* E. Berndt, ed., Chicago, IL: National Bureau of Economic Research, forthcoming.

[3]These expectations are reflected in the International Technology Roadmap for Semiconductors. Accessed at <*http://public.itrs.net/*>.

work—in effect making Moore's Law a self-fulfilling prophecy.[4] For Moore's Law be self-fulfilling, competitors need to believe that potential technological "showstoppers" and other impediments can be effectively overcome in the near to intermediate term. While experts predict that it is possible to remain on the trajectory envisioned by Moore's Law for another 10 to 15 years, this outcome is not inevitable.[5]

Sustaining Moore's Law is important because the production and use of semiconductors are major contributors to the growth and dynamism of the U.S. economy. Semiconductor technologies underpin a variety of products ranging from electronic devices such as personal computers and mobile phones, to business solutions and services, to e-commerce through the Internet.[6] Through their pervasive use and rapid improvement, semiconductors have become technological enablers, allowing major improvements in established products as well as new innovations from consumer electronics (like the iPod) to new medical technologies, to new business processes.

The Committee's focus on semiconductors is based on the central role semiconductors have played and continue to play in the rapid development and better performance of information technologies. Semiconductors are not only a key driver of the performance improvements in information technology, they are unusual in that these performance improvements continue even as costs of computers and other devices keep declining. This in turn has had a positive impact (though one that is hard to measure) on productivity in software development.

To sustain benefits of the new, more productive economy, the Committee recommends a number of policy measures. They include:

Retaining a Vibrant U.S. Information Technology Industry

The structure of the semiconductor, computer, and software industries is changing, with some production as well as advanced R&D moving offshore, creating new opportunities but also new challenges for U.S. leadership. Globalization clearly offers many benefits such as 24/7 product development, high-quality and lower-cost R&D, and lower-cost manufacturing of components and final products. Yet, as more manufacturing and related research and development move outside the United States, the United States risks losing the critical mass necessary for its

[4]Ana Aizcorbe, "Moore's Law, Competition, and Intel's Productivity in the Mid-1990s," *American Economic Review,* 95:305-308, 2005.

[5]See remarks by Robert Doering, "Physical Limits of Silicon CMOS Semiconductor Roadmap Predictions," in National Research Council, *Productivity and Cyclicality in Semiconductors: Trends, Implications, and Questions,* Dale W. Jorgenson and Charles W. Wessner, eds., Washington, D.C.: The National Academies Press, 2004.

[6]European Semiconductor Industry Association, *The European Semiconductor Industry 2005 Competitiveness Report*. Accessed at <*http://www.eeca.org/pdf/final_comp_report.pdf*>.

leadership and autonomy in semiconductor and other information technologies and equipment.[7]

It is important to recognize that the movement of the high-technology industries offshore is not uniquely the result of market forces. While semiconductors and other information technologies are produced and traded in a globally integrated market, national policies often shape international competition. National policies play a major role in attracting investment (e.g., through major tax and R&D incentives) and by creating positive conditions to attract and retain such investment.

The policies of other nations and regions may well pose challenges to U.S. leadership in the high-technology sector.[8] While it is neither possible nor desirable to freeze the allocation of global production, if the United States is to participate successfully in this competition, the federal agencies and state governments will need to undertake measures that strengthen the attractiveness of the United States as a location for the semiconductor, software, and other high-technology research and production, including renewed attention to encouraging and retaining a capable high-tech workforce.[9]

Expanding Research Funding

To sustain the technology trajectory envisaged by Moore's Law requires advanced research to overcome emerging technological "brick walls." Substantial public funding in semiconductor research is necessary if we are to continue to reap the benefits of remaining on the trajectory set out by Moore's Law and for the United States to remain a robust global center for the research, development, and production of semiconductors. To maintain the innovative pace of the semiconductor industry, with the attendant benefits for the U.S. economy, national investments in university research programs that explore and develop promising technologies are needed. Additional government investments in university research for programs that support and move promising technologies closer to

---

[7]President's Council of Advisors on Science and Technology, *Sustaining the Nation's Innovation Ecosystems*, Washington, D.C., 2004, pp. 9 and 14.

[8]National Research Council, *Securing the Future: Regional and National Programs to Support the Semiconductor Industry*, Charles W. Wessner, ed., Washington, D.C.: The National Academies Press, 2003.

[9]George Scalise, "Industry Perspective on Semiconductors," in National Research Council, *Securing the Future: Regional and National Programs to Support the Semiconductor Industry*, op. cit., pp. 35-42. Google's Wayne Rosing reiterated this need at the Committee's workshop on software. See National Research Council, *Software Growth, and the Future of the U.S. Economy*, Dale Jorgenson and Charles Wessner, eds., Washington, D.C.: The National Academies Press, 2006. See also remarks by Craig Barrett at a Semiconductor Industry Association event commemorating the 40th anniversary of Moore's Law. Accessed news release at <*http://www.sia-online.org/pre_release.cfm?ID=355*>.

commercialization are increasingly important to maintain the innovative pace of the information technology industry.[10]

Investing in a Trained Workforce

National investments in a trained workforce that is well grounded in the relevant disciplines—especially physics, chemistry, mathematics, computer science, and engineering—is necessary for world-class research and manufacturing in the semiconductor, computer component, and software industries.[11] Developing the basis for a better-trained workforce begins with strengthening K-12 education. At the secondary level and beyond, scholarships are needed to attract more U.S. students, including women and minorities, to pursue training in computer science and related fields.[12]

In addition to fostering home-born talent, continual progress is also necessary in visa processing in order to attract and retain qualified foreign engineers and scientists. This includes increases in the number of H-1B visas; automatic visa extensions for international students who receive advanced degrees in science, technology, engineering, mathematics, and other fields of national need from U.S. institutions; as well as more permanent opportunities for science and engineering graduates to remain and contribute to the United States economy.[13] Compensation

---

[10]Previous analysis by the Board on Science, Technology, and Economic Policy (STEP) has underscored the importance of innovation partnerships such as the Small Business Innovation Research (SBIR) Program and the Advanced Technology Program in contributing to the development of new technologies. See National Research Council, *The Advanced Technology Program: Assessing Outcomes*, Charles W. Wessner, ed., Washington, D.C.: National Academy Press, 2001, p. 39. See also National Research Council, *The Small Business Innovation Research Program: An Assessment of the Department of Defense Fast Track Initiative,* Charles W. Wessner, ed., Washington, D.C.: National Academy Press, 2000.

[11]Previous analysis by the STEP Board of trends in federal research funding found that "there has been a significant reduction in federal funding in certain of the physical science and engineering fields. These include fields whose earlier advances contributed to the surge in productivity and economic growth of the late 1990s and fields that underlie progress in energy production and conservation, pollution abatement, medical diagnosis and treatment, and other national priorities." National Research Council, *Trends in Federal Support of Research and Graduate Education,* Stephen A. Merrill, ed., Washington, D.C.: National Academy Press, 2001.

[12]See the recommendations in the recent report by the National Academy of Sciences, National Academy of Engineering, and the Institute of Medicine, *Rising Above the Gathering Storm: Energizing and Employing America for a Brighter Future,* Washington, D.C.: The National Academies Press, 2007 Forthcoming, Chapter 5. Legislation reflecting these recommendations is now pending before Congress.

[13]The need for these actions is emphasized in the recent Congressionally mandated NAS/NAE/IOM study, *Rising Above the Gathering Storm: Energizing and Employing America for a Brighter Economic Future,* op. cit. See in particular Actions C-4 through C-6, which call for continuing improvements in visa processing for international students, providing a one-year automatic visa extension to international students who receive doctorates or the equivalent in fields of national need, and the institution of a new skills-based, preferential immigration option that would significantly raise the chances of an applicant with doctoral-level education in science and engineering.

packages for technology workers are also a factor for remaining competitive in attracting and retaining qualified scientists and engineers in the globally competitive information technology industry.[14]

Fostering Public-Private Partnerships

Public-private partnerships, involving cooperative research and development activities among industry, universities, and government laboratories can play an instrumental role in accelerating the development of new technologies and products.[15] Semiconductor industry leaders believe that such partnerships provide the most promising strategy for sustaining Moore's Law, given that the semiconductor industry's ability to make smaller, faster, and cheaper integrated circuits is limited by the growing inability of each firm to pay on its own for the highly expensive research needed to achieve needed innovations.[16]

In addition to pre-competitive research partnerships at the horizontal level (e.g., among semiconductor device manufacturers), vertical partnerships focused on integrated capacities along the supply chain are seen as increasingly important. The objective of the vertical partnerships is to ensure competitiveness across the development and production chain through synergistic relations among suppliers, manufacturers, and users of new advanced technologies.[17]

Developing Industry Roadmaps

Wider adoption of road-mapping exercises by the computer and computer component industries (along the lines of the pre-competitive research charted by the International Technology Roadmap for Semiconductors) can contribute to the ability of information technology industries to remain on a rapid growth

[14]National Research Council, *Building a Workforce for the Information Economy*, Washington, D.C.: National Academy Press, 2001, pp. 69-79.

[15]National Research Council, *Government-Industry Partnerships for the Development of New Technologies: Summary Report*, Charles W. Wessner, ed., Washington, D.C.: The National Academies Press, 2003.

[16]A recent study by SEMI estimates that research required for continued scaling of integrated circuit devices, even without another wafer size increase, will cost some $16.2 billion by 2010. However, the equipment and materials suppliers, to whom the burden of research has shifted from chipmakers, are predicted to be able to afford an annual R&D budget of $10.4 billion, creating a $6 billion gap. SEMI, "Semiconductor Equipment and Materials: Funding the Future," October 2005. Accessed at <*http://content.semi.org/cms/groups/public/documents/homepervasive/p036611.pdf*>.

See also Phil LoPiccolo, "The Six Billion Dollar Gap," *Solid State Technology*, February 2006; and Robert Haavind, "Chipmaking's Tough Economic Road Ahead." *Solid State Technology*, March 2006.

[17]European Semiconductor Industry Association, *The European Semiconductor Industry: 2005 Competitiveness Report,* op. cit., Executive Summary, p. 51.

path, like that predicted by Moore's Law, accelerating the pace of innovation and growth.[18]

Setting International Standards

The economic stakes of standard setting are of great consequence. The use of relevant technical standards, such as for software and wireless devices, is an important element in sustaining the success of the United States in the global economy. Conversely, uncertainty created by a multiplicity of standards and a lack of clarity in regulatory policy can retard progress, as seen in the U.S. broadband gap.[19] Some nations and regions see standards as a competitive tool and devote substantial resources to this end. The role and resources of the National Institute of Standards and Technology have to be seen in this light. The standard-making process must be recognized as a key component of U.S. competitiveness and provided commensurate resources and policy attention.

Revising Outdated Telecom Regulation

Although massive investments in the nation's high-capacity Internet backbone have created excess capacity in long-haul facilities, a variety of factors—regulation among them—have slowed the build-out of the crucial last mile. Indeed, by creating highly technology-specific industry rules and by attempting to promote competition by requiring incumbents to share the local loops of their network with rivals, the Telecommunications Act of 1996 may have, according to some experts, inadvertently inhibited investment needed to provide high-bandwidth access over the last mile.[20] This broadband bottleneck inhibits a fuller capitalization of substantial investments in information technology (IT) and infrastructure, limiting the potential for sustained growth in the economy. Some fear that over time, this slow pace of broadband build-out may result in a competitive disadvantage for the United States, although emerging wireless technological standards

---

[18]An example is the roadmap exercise by the U.S. Display Consortium, which develops platform technologies for flat-panel displays. See <*http://www.usdc.org/*> for additional information.

[19]See Action D-4 on ensuring ubiquitous broadband Internet access in NAS/NAE/IOM, *Rising Above the Gathering Storm: Energizing and Employing America for a Brighter Future*, op. cit.

[20]Robert Litan and Roger G. Noll, "The Uncertain Future of the Telecommunications Industry," Brookings Working Paper, Washington, D.C.: The Brookings Institution, December 3, 2003. Interpretations vary on the impact of the 1996 Telecommunications Act. Some experts believe that competition for the provision of broadband was already taking place in most major downtown areas in many of the largest cities of the United States before the Telecommunications Act. See Glenn Woroch, "Local Network Competition," in *Handbook of Telecommunications Economics*, Martin Cave, Sumit Majumdar, and Ingo Vogelsang, eds., New York, NY: Elsevier, 2002. Others believe that the act did not deter the build-out of the nation's cable network. For example, see Jonathan E. Nuechterlein and Philip J. Weiser, *Digital Crossroads: American Telecommunications Policy in the Internet Age,* Cambridge, MA: The MIT Press, 2005.

such as WiMAX may help overcome some of the limitations associated with traditional wired broadband.[21]

Developing a New Architecture for U.S. National Accounts

Originally constructed to deal with economic stabilization arising from the Great Depression of the 1930s, the basic architecture of the national accounts has not been substantially altered in 50 years. In the meantime, the policy focus has shifted from stabilization of the economy to enhancing the economy's growth potential. The U.S. national accounts require a new architecture to inform policy makers confronting new challenges arising from rapid changes in technology and globalization. Additional resources should be made available to further explore this call for a new architecture. The drivers of the U.S. economy have evolved, indeed shifted quite dramatically, and it is essential that a new architecture for the national accounts be put into place to better capture this new reality.

MEASURING THE NEW ECONOMY

Given the benefits of rapid technical innovation, the measurement issues associated with this change should be addressed on a systematic basis by the responsible agencies of the federal government in a coordinated fashion. Swiftly falling IT prices provide powerful economic incentives for the diffusion of information technology. Given that the rate of the IT price decline is a key component of the cost of capital, it is essential to develop constant quality indexes, such as those for computers, for use in the U.S. National Income and Product Accounts (NIPA). The development of quality-adjusted prices is necessary for all types of investment, not just IT investment. Other types of investment have also benefited from technological progress and a failure to implement quality adjustments for all products will eventually lead to biased statistics.

Additional resources to develop price indexes and related analyses are needed to understand the sources of productivity growth in the economy and to develop the policies to sustain it.

Developing a Forecasting Model for the Semiconductor Industry

Although information technology is altering product markets and business organizations, a fully satisfactory model of the semiconductor industry remains to be developed. Such a model would derive the demand for semiconductors from investments in information technology in response to rapidly falling IT prices. An

[21]For a variety of views on closing the putative broadband gap, see National Research Council, *The Telecommunications Challenge: Changing Technologies and Evolving Policies*, Charles W. Wessner, ed., Washington, D.C.: The National Academies Press, 2006.

important objective is to determine the product cycle for successive generations of new semiconductors endogenously.

Developing Constant Quality Price Indexes for Computers and Computer Components

Economists require accurate measures of the performance of computers and computer components in order to understand their contributions to economic growth. The computer component industry has developed a variety of formal and informal measures to gauge the relative performance of its products. Further development of these measures (using the Hedonic method[22]) and subsequent incorporation into the National Income and Product Accounts should enable improved analysis and policies to sustain the contributions of computers and computer components to economic growth.

Developing Constant Quality Price Indexes for Software

Software price indexes, especially for own-account and custom software, must be upgraded to hold software performance constant. Without adjustment for quality, these indexes present a distorted picture of software prices as well as software output and investment. To this end, advances in developing software price indexes, including current work by the Bureau of Economic Analysis (BEA) on function points, hedonic techniques, and other methodologies, should be supported.[23] These advances can improve statistical information on firm investments in customized software applications such as own-account and custom

[22]Hedonic price indexes provide a proven method for adjusting for quality differences in computers across time. Using this method requires improved performance measures for computers and computer components. Gregory Chow pioneered the use of hedonic techniques for constructing a constant quality index of computer prices in research conducted at IBM. See Gregory C. Chow, "Technological Change and the Demand for Computers," *American Economic Review*, 57(5):117-130, December 1967. In 1985, BEA incorporated constant quality price indexes for computers and peripheral equipment constructed by IBM into the National Income and Product Accounts (NIPA). The economic interpretation of these indexes by Jack Triplett brought the rapid decline of computer prices to the attention of a very broad audience. See Jack Triplett, "The Economic Interpretation of Hedonic Methods," *Survey of Current Business*, 66(1):36-40, January 1986. Triplett has also provided exhaustive surveys of research on hedonic price indexes for computers. See, for example, Jack Triplett, *Handbook on Hedonic Indexes and Quality Adjustments in Price Indexes: Special Application to Information Technology Products*. Paris: Organisation for Economic Co-operation and Development, 2004.

[23]BEA has recently launched a software pricing project for custom and own-account software using function points. Work by Q/P Management Group and the Analysis Group is expected to produce new price indexes for custom and own-account software for the U.S. national accounts. A function point metric is a means of measuring software size and productivity. It uses functional, logical entities such as inputs, outputs, and inquiries that tend to relate more closely to the functions performed by the software. See John J. Marciniak, ed., *Encyclopedia of Software Engineering,* New York, NY: John Wiley & Sons, pp. 518-524, 1994.

software.[24] Furthermore, the adoption of common standards across the Organisation for Economic Co-operation and Development (OECD) and beyond should also be encouraged. Wider use of standards can improve our knowledge about investments in software in what is a global industry and facilitate the tracking of software outsourcing.[25]

Developing Constant Quality Price Indexes for Telecommunications

The varying complexity and rates of technical innovation make the contribution of telecommunications equipment to productivity growth a challenge to measure. Current BEA methodologies for making inter-temporal comparisons in price and quality understate true price declines in communications equipment because they do not fully track evolving technological changes.[26] While the Producer Price Index has tried to address some of these changes using hedonic techniques, data that consistently identify important current period product characteristics and transaction prices are not yet readily available.[27] Research into alternative quality valuation techniques and improved data transparency is required to respond to the technological changes in telecommunications equipment. BEA and other statistical agencies require increased funding to follow evolving trends in the communications arena with more accuracy.

Gauging the Scope of Globalization

Although the offshoring phenomenon—particularly the offshoring of service-sector jobs—is a topic of much currency, the scope of the phenomenon remains to be adequately documented. Despite the media attention, there is relatively little hard information about the causes and impact of offshoring on manufacturing and service-sector employment in the United States or on other related economic and structural developments. A sustained effort to measure the dimensions and implications of offshoring is necessary for informed policy making. The necessary resources should be made available to provide better information both to policy makers and to the general public.

---

[24]David Wasshausen, "A BEA Perspective: Private Fixed Software Investment," in National Research Council, *Software, Growth, and the Future of the U.S. Economy,* op. cit.

[25]See comments by Dirk Pilat, "What is in the OECD Accounts and How Good is it?" in National Research Council, *Software, Growth, and the Future of the U.S. Economy,* op. cit.

[26]BEA estimated that prices for communications gear fell an average of 3.2 percent per year between 1994 and 2000. Recent analysis by Marc Doms however suggests that communications equipment prices actually fell about 8 to 10 percent over that period. Mark Doms, "Communications Equipment: What Has Happened to Prices?" *Federal Reserve Bank of San Francisco Working Paper 2003-15.*

[27]For additional perspective on the types of technological changes in telecom equipment that, at least conceptually, could be valued in a hedonic model, see Michael Holdway, "Confronting the Challenge of Estimating Constant Quality Price Indexes for Telecommunications Equipment in the Producer Price Index," Working Paper, Washington, D.C.: Bureau of Economic Analysis, 2002.

I

INTRODUCTION

The Emergence of the New Economy

Dale W. Jorgenson
Harvard University

The resurgence of the American economy since 1995 has now survived the dot-com crash, the short recession of 2001, and the tragedy of 9/11.[1] The unusual combination of more rapid growth and slower inflation has touched off a strenuous debate about whether improvements in America's economic performance can be sustained. A consensus has emerged that the development and deployment of information technology (IT) is the foundation of the American growth resurgence.[2] The mantra of the "new economy"—faster, better, cheaper—characterizes the speed of technological change and product improvement in semiconductors, the key enabling technology.

In 1965 Gordon Moore, then research director at Fairchild Semiconductor, made a prescient observation, later known as Moore's Law.[3] Plotting data on integrated circuits, he observed that each new device contained roughly twice as many transistors as the previous one and was released within 12-24 months of its predecessor. This implied exponential growth of chip capacity at 25-50 percent per year! Moore's Law, formulated in the infancy of the semiconductor industry,

[1]Jorgenson, Ho, and Stiroh (2004) present projections of U.S. economic growth. See Dale W. Jorgenson, Mun S. Ho, and Kevin J. Stiroh, "Will the U.S. Productivity Resurgence Continue?" *Federal Reserve Bank of New York Current Issues in Economics and Finance*, 10(13):1-7, 2004. Available at <*http://www.newyorkfed.org/research/current_issues/ci10-13.html*>.

[2]The role of information technology in the American growth resurgence is discussed in detail by Jorgenson, Ho, and Stiroh (2005). See Dale W. Jorgenson, Mun S. Ho, and Kevin J. Stiroh, *Information Technology and the American Growth Resurgence*, Cambridge, MA: The MIT Press, 2005.

[3]See Gordon E. Moore, "Cramming More Components onto Integrated Circuits," *Electronics*, 38(8):114-117, 1965. Available at <*ftp://download.intel.com/research/silicon/moorespaper.pdf*>.

has tracked chip capacity for 40 years. Moore recently extrapolated this trend for at least another decade.[4]

The economics of information technology begins with the precipitous and continuing fall in semiconductor prices. Moore emphasized this price decline in his original formulation of Moore's Law, and dramatically plunging prices are used almost interchangeably with faster and better devices in describing the evolution of semiconductor technology. The rapid price decline has been transmitted to the prices of a range of products that rely heavily on this technology, like computers and telecommunications equipment. The technology has also helped to reduce the costs of aircraft, automobiles, scientific instruments, and a host of other products.

Swiftly falling IT prices provide powerful economic incentives for the rapid diffusion of information technology. A substantial acceleration in the IT price decline occurred in 1995, triggered by a much sharper acceleration in the price decline for semiconductors. This can be traced to a shift in the product cycle from 3 years to 2 years as a consequence of intensifying competition in semiconductor markets. Continuation of this shorter product cycle for the next decade is consistent with the technological developments projected in the most recent International Technology Road Map for Semiconductors.[5]

The accelerated IT price decline since 1995 signals faster productivity growth in IT-producing industries—semiconductors, computers, communications equipment, and software. These industries have accounted for a substantial share of the surge in U.S. economic growth. It is important, however, to emphasize that accelerating growth is not limited to these industries. To analyze the impact of the accelerated price decline in greater detail, it is useful to divide the remaining industries between IT-using industries, those particularly intensive in the utilization of IT equipment and software, and the non-IT industries.

Although three-quarters of U.S. industries have contributed to the acceleration in economic growth, the four IT-producing industries are responsible for a quarter of the growth resurgence, but only 3 percent of the Gross Domestic Product (GDP). IT-using industries account for another quarter of the growth resurgence and about the same proportion of the GDP, while non-IT industries with 70 percent of value-added are responsible for only half the resurgence. Obviously, the impact of the IT-producing industries is far out of proportion to their relatively small size.

In view of the critical importance of productivity, it is essential to define this concept more precisely. Productivity is defined as output per unit of input, where

---

[4]See Gordon E. Moore, "No Exponential is Forever . . . But We Can Delay Forever," International Solid State Circuit Conference, San Francisco, CA, February 10, 2003. Available at *<ftp://download.intel.com/research/silicon/Gordon_Moore_ISSCC_021003.pdf>*.

[5]International SEMATECH, "International Technology Roadmap for Semiconductors," Austin, TX, December 2004. Available at *<http://public.itrs.net/>*.

input includes capital and labor inputs as well as purchased inputs.[6] This definition has the crucial advantage of clearly identifying the role of purchased goods and services, such as semiconductors used by other IT-producing industries. The purchased goods and services are the components of the industry's inputs that are "outsourced" in order to make the most of the advantages of specialization.

Industry inputs consist of capital, labor, and purchased inputs. It is remarkable that four IT-producing sectors taken together have the most rapid growth of all three. The surging growth of the four IT-producing industries has its sources in both inputs and productivity; however, the relative importance of these sources differs considerably. All the IT-producing industries have large contributions of purchased goods and services, including inputs from other IT-producing sectors. The software industry has the most rapidly growing labor input, but almost no productivity growth.

Two industries responsible for much of IT hardware—computers and semiconductors—exhibit truly extraordinary rates of productivity growth, as well as a substantial acceleration in the growth of productivity after 1995. As a group, the four IT-producing industries contribute more to economy-wide productivity growth than all the other industries combined. In fact, the contributions of the IT-using and non-IT industries to the economy's productivity growth have been slightly negative, partly offsetting the positive contribution of the IT-producing industries.

However, investment rather than productivity has been the predominant source of U.S. economic growth throughout the postwar period. The rising contribution of investment since 1995 has been the key contributor to the U.S. growth resurgence and has boosted growth by close to a full percentage point. The contribution of IT investment accounts for more than half of this increase. Investment in computers has been the predominant impetus to faster growth, but communications equipment and software investments have also made important contributions.[7]

Accelerated capital growth reflects the surge of investment in IT equipment and software after 1995 in the large IT-using sectors like Finance and Trade. However, virtually all industries have responded to more rapid declines in IT prices by substituting IT for non-IT capital. Capital from IT products has grown at double-digit rates during most of the last three decades. By contrast, non-IT capital has grown at about the same rate as the economy as a whole, an order of

[6]In economic jargon this definition is often referred to as "total factor productivity." This must be carefully distinguished from the more common "labor productivity," output per hour worked. To avoid confusion I will use the term "productivity" only in the sense of total factor productivity or output per unit of all inputs.

[7]For an explanation of how the relative contributions are measured, see Dale W. Jorgenson, Mun S. Ho, and Kevin J. Stiroh, *Productivity, Volume 3: Information Technology and the American Growth Resurgence*, Cambridge, MA: The MIT Press, 2005.

magnitude more slowly. Half of U.S. industries actually show a declining contribution of non-IT capital.

While the IT-producing industries demonstrate accelerating growth in every dimension, the impact is limited by their relatively small size. IT-using sectors are especially prominent in the accelerated deployment of IT equipment and software, while the non-IT industries contribute impressively to faster productivity growth. After 1995 IT-producing industries show sharply accelerating growth in productivity, while IT-using industries diverge from this trend by exhibiting a more rapid decline. Productivity growth in non-IT industries has jumped very substantially, accounting for much of the acceleration in economy-wide productivity.

The very modest acceleration in employment growth after 1995 has been concentrated in IT-using industries. Since the number of workers available for employment is determined largely by demographic trends, the acceleration in IT investment is reflected in rates of labor compensation and changes in the industry distribution of employment. The rapidly growing IT-using industries have absorbed large numbers of college-educated workers, while non-IT industries have shed substantial numbers of non-college workers.

The surge of IT investment in the United States after 1995 has counterparts in all other industrialized economies. Using "internationally harmonized" IT prices that rely primarily on U.S. trends, the burst of IT investment in all industrialized economies that accompanied the acceleration in the IT price decline in 1995 is revealed unmistakably. These economies have also experienced a rise in productivity growth in the IT-producing industries. However, differences in the relative importance of these industries have generated wide disparities in the impact of IT on economic growth. Among the G7 countries–Canada, France, Germany, Italy, Japan, the United Kingdom, and the United States—the role of the IT-producing industries is greatest in the United States.

To conclude: The mechanism underlying the resurgence of U.S. economic growth has now come into clear focus.[8] The surge was generated by the accelerating decline of IT prices, propelled by a shift in the semiconductor product cycle from 3 years to 2 in 1995. The price decline set off an investment boom that achieved its peak during the last half of the 1990s and has now recovered much of the momentum lost during the 2001 recession. Achievement of the ambitious goals of the International Technology Roadmap for Semiconductors (2004) will greatly help to ensure that the America's improved economic performance can be sustained.

---

[8]More detail on this mechanism is provided by Jorgenson (2001). See Dale W. Jorgenson, *Economic Growth in the Information Age*, Cambridge, MA: The MIT Press, 2001.

II

FINDINGS AND RECOMMENDATIONS

Findings and Recommendations

INTRODUCTION

The findings and recommendations found in this section reflect the Committee's consensus based on its own deliberations as well as on the proceedings of five previous conferences that explored the operation of the new, information-based economy. The Committee's aim is to understand the sources of productivity growth in this new economy, to measure more accurately the contributions of different components of growth, and to develop policies to encourage and increase that growth.

A. THE NATURE OF THE NEW U.S. ECONOMY

Findings

1. The New Economy refers to technological and structural changes in the U.S. economy as individuals capitalize on new technologies, new opportunities, and national investments in computing, information, and communications technologies. These structural changes have resulted in a long-term positive productivity shift of major significance.[1]

[1]Dale W. Jorgenson and Kevin J. Stiroh, "Raising the Speed Limit: Economic Growth in the Information Age," in National Research Council, *Measuring and Sustaining the New Economy*, Dale W. Jorgenson and Charles W. Wessner, eds., Washington, D.C.: National Academy Press, 2002, Appendix A.

a. Despite differences in methodology and data sources, a consensus has emerged among economists that the remarkable behavior of information technology (IT) prices provides the key to the surge in U.S. economic growth after 1995.
b. The relentless decline in the prices of information technology equipment has steadily enhanced the role of IT investment across the economy.[2] Productivity growth in IT-producing industries has risen in importance and a productivity revival is under way in the rest of the economy.
c. The decline in IT prices refers to more than just a reduction in the price of a key economic input. The widespread use of IT, made possible by this price reduction, has changed and continues to change how individuals and businesses in the economy work, consume, communicate, and transact. New products and capabilities made possible by lower-cost computing and communications facilities are already restructuring the economy and accelerating the globalization of manufacturing and trade in services, with major positive implications for productivity growth.
d. New information technologies have a broad and positive impact on U.S. productivity growth through industries that produce new information technologies and the many more that apply them. New IT applications are also contributing to enhanced workplace productivity as a wide variety of firms adapt to changes in information flows and take advantage of new organizational structures made possible by these innovations.[3] These developments are changing the structure of firms, creating more innovative and more agile enterprises, with positive indirect and long-term implications for productivity growth.[4]

[2]Kevin Stiroh notes that over the last decade, U.S. firms have invested over $2.4 trillion in IT assets, such as computer hardware, computer software, and telecommunications equipment and that these three assets accounted for more than 40 percent of private fixed investment in equipment and software in 2000. Kevin J. Stiroh, "Measuring Information Technology and Productivity in the New Economy," *World Economics* 3(1):43-59, 2002.

[3]These indirect effects are captured in a now substantial literature. For example, see Sandra E. Black and Lisa M. Lynch, "What's Driving the New Economy?: The Benefits of Workplace Innovation," *Economic Journal*, 114(493):F97-F116, 2004. See also Timothy Bresnahan, Erik Brynjolfsson and Lorin M. Hitt, "Information Technology, Workplace Organization and the Demand for Skilled Labor: Firm-level Evidence," *Quarterly Journal of Economics*, 117(1):339-376, 2002. For a discussion of how IT interacts with other aspects of firm structure, labor policies, and innovation such as human capital, improved organizational structure, and incentives, see Erik Brynjolfsson and Lorin M. Hitt, "Computing Productivity: Firm-Level Evidence," *Review of Economics and Statistics,* 85(4):793-808, 2003.

[4]See Amar Bhidé, "Venturesome Consumption, Innovation, and Globalization," Paper prepared for a joint conference of CESIFO and the Center on Capitalism and Society on "Perspectives on the Performance of the Continent's Economies," Venice, July 21-22, 2006. Bhidé notes that an important part of innovation centers on the incentives facing firms and individuals in trying new products and reorganizing themselves to take advantage of new products.

2. Cheaper information technology has given greater importance to more productive forms of capital. The rising contribution of investments in information technology since 1995 has been a key contributor to the U.S. growth resurgence and has boosted growth by close to a percentage point.
   a. The contribution of investment in information technology accounts for more than half of this increase. Within information technology, computers have been the predominant impetus for faster growth. Communications equipment and software have also made important contributions to growth.
   b. Altogether, 31 industries (out of the 44 industry categories that make up the U.S. economy) contributed to the acceleration in economic growth after 1995. The four IT-producing industries discussed here are responsible for only 2.9 percent of the Gross Domestic Product (GDP) but a remarkable quarter of the U.S. growth resurgence.[5] The 17 IT-using industries account for another quarter of the surge in growth and about the same proportion of the GDP, while the non-IT industries with 70 percent of value added are responsible for half the resurgence. The contribution of the IT-producing industries is far out of proportion to their relatively small size in relation to the economy as a whole. These industries have grown at double-digit rates throughout the period 1977-2000, but their growth jumps sharply after 1995, when the GDP share of these industries also increases.[6]
   c. The accelerated IT price decline also signals faster total factor productivity growth in IT-producing industries.[7] The four IT-producing industries contributed more to the growth of total factor productivity during the period 1977-2000 than all other industries combined.[8]

[5]Dale W. Jorgenson, Mun S. Ho, and Kevin J. Stiroh, *Productivity, Volume 3: Information Technology and the American Growth Resurgence*, Cambridge, MA: The MIT Press, 2005, p. 10.

[6]Ibid., p. 11.

[7]Total factor productivity is defined as output per unit of input, where input includes capital, labor, and intermediate inputs. There is some debate among economists about how easy it is to infer growth rates of total factor productivity growth from relative price declines. While Aizcorbe concludes that quality change was the dominant source of the increase in relative price declines in the mid-1990s, others might disagree. (See Ana Aizcorbe, "Why Are Semiconductor Prices Falling so Fast? Industry Estimates and Implications for Productivity Growth," *Economic Inquiry*, forthcoming.) For example, Hobijn argues that declining margins cloud this linkage and Aizcorbe presents a model where increased competition leads to accelerated price declines. See Bart Hobijn, "Is Equipment Price Deflation a Statistical Artifact?" *Federal Reserve Bank of New York Staff Report #139*, November 2001. Also see Ana Aizcorbe, "Moore's Law, Competition, and Intel's Productivity in the Mid 1990s," *American Economic Review*, 95:305-308, May 2005.

[8]Dale W. Jorgenson, Mun S. Ho, and Kevin J. Stiroh, *Productivity, Volume 3: Information Technology and the American Growth Resurgence*, op. cit.

3. Gains in the U.S. terms of trade, especially for information technology products, may have contributed to the acceleration in U.S. productivity in the late 1990s.[9]
   a. Information technology is one of the most globally engaged sectors of the U.S. economy. Liberalization of information technology trade began in the 1980s, helping to decrease the cost of semiconductors and increase the availability of IT products and services.[10] The International Technology Agreement of 1996 also eliminated all world tariffs on hundreds of IT products in four stages from early 1997 through 2000, helping to lower the prices of imported intermediate IT products.[11]
   b. While such trade effects are likely to explain only a small portion of the productivity speed-up, foreign trade practices do appear to matter for the measurement of productivity.

4. Improved productivity associated with the introduction of advanced information and communications technologies appears to have raised the long-term growth trajectory of the U.S. economy. This gain appears to be robust, having survived the dot-com crash, the short recession of 2001, and the tragedy of 9/11. Since the end of the previous recession in 2001, productivity growth has been running at about two-tenths of a percentage point higher than in any recovery of the post-World War II period.[12]

5. A structural change most associated with the New Economy today is the transformation of the Internet from a communication media to a platform for service delivery.[13] This has contributed to the remarkable growth of the U.S. service

---

[9]Robert C. Feenstra, Marshall B. Reinsdorf, and Michael Harper, "Terms of Trade Gains and U.S. Productivity Growth," paper prepared for NBER-CRIW Conference, July 25, 2005.

[10]For a detailed analysis of trade in semiconductors, see Kenneth Flamm, *Mismanaged Trade*, Washington, D.C.: Brookings Institution Press, 1996. For a review of the impact of the 1986 semiconductor trade agreement on the revival of the U.S. semiconductor industry, see National Research Council, *Securing the Future: Regional and National Programs to Support the Semiconductor Industry,* Charles W. Wessner, ed., Washington, D.C.: The National Academies Press, 2003, p. 82. The report points out that the resurgence of the U.S. semiconductor industry was based in part on the success of the SEMATECH consortium, in part on the 1986 Semiconductor Trade Agreement, and in part on the repositioning of the U.S. industry away from DRAM chips and towards microprocessor design and production. The recovery of the U.S. industry was thus like a three-legged stool; it is unlikely that any one factor would have proved sufficient independently.

[11]For an overview of the Information Technology Agreement and its implementation, access the World Trade Organization Web site at <*http://www.wto.org/english/tratop_e/inftec_e/itaintro_e.htm*>.

[12]Dale W. Jorgenson, Mun S. Ho, and Kevin J. Stiroh, "Will the U.S. Productivity Resurgence Continue?" *Federal Reserve Bank of New York Current Issues in Economics and Finance,* 10(13), 2004.

[13]This transformation is sometimes referred to as "Web 2.0." For a description of this new version of the Web, see Tim O'Reilly, "What Is Web 2.0—Design Patterns and Business Models for the Next Generation of Software" September 30, 2005. Accessed at <*http://www.oreillynet.com/pub/a/oreilly/tim/news/2005/09/30/what-is-web-20.html*>.

economy, as companies like Google and eBay increasingly exploit information services in new ways. As new business models, enabled by the Web, continue to emerge, they will contribute to sustaining the productivity growth of U.S. economy.

Recommendations

1. Given the benefits of rapid technical innovation, the measurement issues associated with this change should be addressed on a systematic basis by the responsible agencies of the federal government in a coordinated fashion.
   a. Swiftly falling IT prices provide powerful economic incentives for the diffusion of information technology. Given that the rate of the IT price decline is a key component of the cost of capital, it is essential to develop constant quality indexes, such as those for computers, for use in the U.S. National Income and Product Accounts (NIPA).
   b. Substantial resources to develop price indexes and related analyses are needed to understand the sources of productivity growth in the economy and to develop the policies to sustain it.

2. The growing synergies and new economic opportunities of the New Economy need to be understood better if they are to be sustained through appropriate policies. The rapid pace of these changes means that they require regular and systematic monitoring in order to bring significant changes to the attention of policy makers.
   a. The rapid business and workplace transformations made possible by information technology are not only a product of globalization but also a factor that is advancing globalization.[14] For the United States, success in this new global paradigm requires technological leadership as well as strategic use of information technology.
   b. To remain a leader in information and communications technologies, the United States must foster and attract the best human resources. Both the federal and state governments must also adequately support research funding, and maintain a superior business environment and encourage the public-private partnerships that foster innovation and the timely transition of research to the marketplace.[15] It must also update regulations that inhibit wider access to and use of information networks.[16]

[14]Catherine L. Mann, *High-Technology and the Globalization of America*, forthcoming.

[15]National Research Council, *Government-Industry Partnerships for the Development of New Technologies: Summary Report,* Charles W. Wessner, ed., Washington, D.C.: The National Academies Press, 2003.

[16]National Academy of Sciences, National Academy of Engineering, and Institute of Medicine, *Rising Above the Gathering Storm: Energizing and Employing America for a Brighter Economic Future,* Washington, D.C.: The National Academies Press, 2007 Forthcoming.

B. MOORE'S LAW AND THE NEW ECONOMY

Findings

1. Faster and cheaper semiconductors are a key driver of the productivity gains associated with the recent growth of the U.S. economy.[17]
   a. Price-performance improvement in semiconductors has been a major source of price-performance improvement in information technology. Declines in cost for electronics functionality embedded in semiconductors are the linchpin of improvement in price-performance for computers and communications, which in turn has been a major factor in the increase in long-term growth performance.[18]
   b. A substantial acceleration in the pace of IT price decline occurred in 1995, triggered by a much sharper acceleration in the price decline of semiconductors—the key component of modern information technology.[19] (See Figure 1.[20]) This acceleration can be traced to a shift in the product cycle from 3 years to 2 years as a result of intensifying competition in markets for semiconductor products.[21]

---

[17]Dale W. Jorgenson, Mun S. Ho, and Kevin J. Stiroh, *Productivity, Volume 3: Information Technology and the American Growth Resurgence,* op. cit.

[18]Jack E. Triplett, "High-Tech Productivity and Hedonic Price Indexes," in Organisation for Economic Co-operation and Development, *Industry Productivity*, Paris: Organisation for Economic Cooperation and Development, 1996; Kenneth Flamm, "Technological Advance and Costs: Computers vs. Communications," in *Changing the Rules: Technological Change, International Competition, and Regulation in Communications*, Robert C. Crandall and Kenneth Flamm, eds., Washington, D.C.: The Brookings Institution, 1989; Ana Aizcorbe, Kenneth Flamm, and Anjum Khurshid, "The Role of Semiconductor Inputs in IT Hardware Price Declines" in *Hard to Measure Goods and Services: Essays in Honor of Zvi Griliches*, E. Berndt, ed., Chicago, IL: National Bureau of Economic Research, forthcoming.

[19]Using industry estimates on Intel's operations to decompose a price index, Ana Aizcorbe finds that virtually all of the declines in a price index for Intel's chips can be attributed to quality increases associated with product innovation, rather than declines in the cost per chip. She adds that consistent with the inflection point that Jorgenson noted in the overall price index for semiconductors, the Intel price index falls faster after 1995 than in the earlier period, but that the decomposition attributes virtually all of the inflection point to an acceleration in quality increases. These increases in quality push down constant quality costs. See Ana Aizcorbe, "Why Are Semiconductor Price Indexes Falling So Fast? Industry Estimates and Implications for Productivity Growth," op. cit. See also Dale W. Jorgenson, "Information Technology and the U.S. Economy," *American Economic Review,* 91(1), 2001.

[20]The output price index referred to in Figure 1 is the GDP deflator, but differs from the typical Bureau of Economic Analysis (BEA) GDP deflator due to methodology. We impute a capital service flow for government and consumer durable capital and use Tornqvist aggregation to add components of GDP.

[21]For an analysis of the break points in prices of microprocessors, see Ana Aizcorbe, Stephen D. Oliner, and Daniel E. Sichel, "Shifting Trends in Semiconductor Prices and the Pace of Technological Progress," mimeo, Federal Reserve Board, April. Other analyses focus on an acceleration in the pace of technological innovation in semiconductor manufacturing as accelerating the decline in prices: see Kenneth Flamm, "Microelectronics Innovation: Understanding Moore's Law and Semiconductor

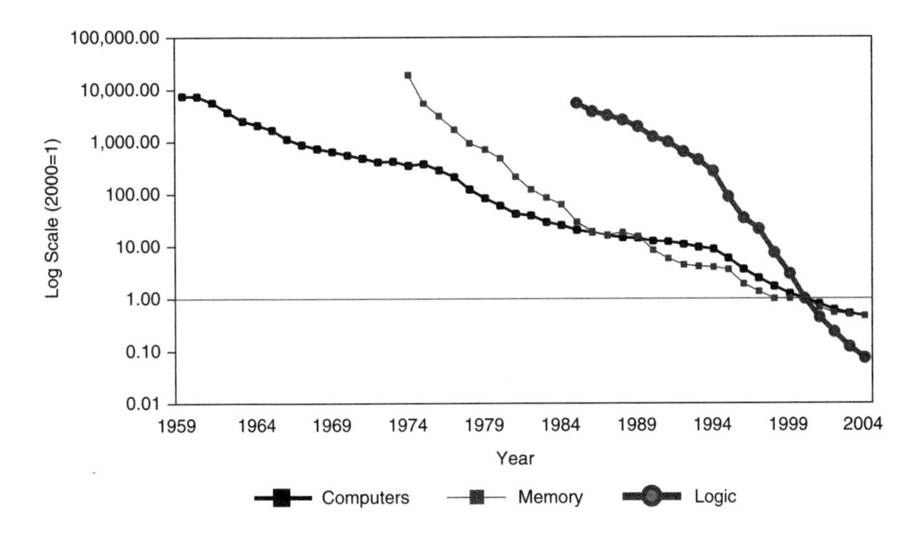

FIGURE 1 Relative prices of computers and semiconductors, 1959-2004.
NOTE: All price indexes are divided by the output price index.

c. Although the decline in semiconductor prices has been projected to continue for at least another decade, the magnitude of recent acceleration may be temporary.

2. Moore's Law has played a significant role in the expectations and development of the semiconductor industry. While by no means dictating an actual law, Gordon Moore correctly foresaw in 1965 the rapid doubling of the feature density of a chip, now interpreted as approximately every 18 months.[22]
   a. While not pretending to be deterministic, Moore's formulation has endured in part by setting expectations among participants in the semiconductor industry of the pace of innovation and the introduction of

Price Trends," *International Journal of Technology, Policy, and Management*, 3(2), 2003; Kenneth Flamm, "The New Economy in Historical Perspective: Evolution of Digital Technology," in *New Economy Handbook*, Derek C. Jones, ed., Academic Press, 2003; and Kenneth Flamm, "Moore's Law and the Economics of Semiconductor Price Trends," in Dale W. Jorgenson and Charles W. Wessner, eds., *Productivity and Cyclicality in Semiconductors: Trends, Implications, and Questions*, Washington, D.C.: The National Academies Press, 2004.

[22]Observing that the number of transistors per square inch on integrated circuits had doubled every year since the integrated circuit was invented, Gordon Moore predicted in 1965 that this trend would continue for the near future. (See Gordon E. Moore, "Cramming More Components onto Integrated Circuits," *Electronics*, 38(8), April 19, 1965.) The current definition of Moore's Law, which has been acknowledged by Dr. Moore, holds that the data density of a chip will double approximately every 18 months. Many experts expect Moore's Law to hold for another 15 years.

new products to market.[23] Each firm believes that its competitors will release the next model in an 18-month timeframe, leading each to set the pace of its own work on this basis—in effect making Moore's Law a self-fulfilling prophecy.[24] Currently, the industry expects to remain on the trajectory envisioned by Moore's Law for another 10 to 15 years.[25]

b. Making additional assumptions, an economic corollary to Moore's Law is a rapid fall in the relative prices of semiconductors. With the acceleration in manufacturing innovation in the late 1990s came an increase in the rate of price decline—from roughly 15 percent annually in the early 1990s to 28 percent annually after 1995 until 2003. The increase in chip capacity and the concurrent fall in price—the "faster-cheaper" effect—have created powerful incentives for firms to substitute information technology for other forms of capital. These investments, when effectively integrated, have led to the productivity increases that are the hallmark of the phenomenon known as the New Economy.[26]

3. The Semiconductor Industry Roadmap has helped to sustain Moore's Law.
   a. The International Technology Roadmap for Semiconductors (ITRS) helps set the competitive pace of the semiconductor industry. By identifying common research challenges and reducing costs by identifying redundancies and technical "showstoppers," the ITRS process helps the semiconductor industry commit to the investments necessary to stay on the growth trajectory of Moore's Law.[27]
   b. In 1997, the ITRS reported the presence of a faster two-year semiconductor cycle beginning in 1995 that has helped to accelerate the pace of Moore's Law. However, the 2003 edition of the ITRS has predicted that (given the difficulties encountered at the 90 nm technology node among other reasons) chipmakers will soon return to a 3-year cycle between technology nodes, significantly slowing the pace of semiconductor development.[28]

---

[23]These expectations are reflected in the International Technology Roadmap for Semiconductors. Accessed at <*http://public.itrs.net/*>.

[24]See Kenneth Flamm, "Moore's Law and the Economics of Semiconductor Price Trends," op. cit., 2004, and "Microelectronics Innovation: Understanding Moore's Law and Semiconductor Price Trends," op. cit., 2003. See also Ana Aizcorbe, "Moore's Law, Competition, and Intel's Productivity in the Mid-1990s," BEA Working Paper WP2005-8, September 1, 2005.

[25]See remarks by Robert Doering, "Physical Limits of Silicon CMOS Semiconductor Roadmap Predictions," in National Research Council, *Productivity and Cyclicality in Semiconductors: Trends, Implications, and Questions,* op. cit.

[26]Dale W. Jorgenson, "Information Technology and the U.S. Economy," op. cit.

[27]William Spencer, Linda Wilson, and Robert Doering, "The Semiconductor Technology Roadmap," *Future Fab International*, 18, January 12, 2005.

[28]Access the ITRS homepage at <*http://public.itrs.net/*>.

4. The semiconductor industry is characterized by high annual growth averaging around 15 percent per annum.[29] This high growth rate is accompanied by considerable market volatility, reflected in significant cyclical swings in production.
    a. In large part, this volatility is because the semiconductor industry is highly capital-intensive, requiring significant capital expenditures for each fabrication facility and a very high intensity of R&D (sometimes up to 20 percent of revenue). This high level of investment underpins the high rate of innovation evident through increased performance, miniaturization, cost reduction, and short design cycles.[30]
    b. High sunk costs, steep learning curves, and rapid shifts in product cycles all contribute to a high level of industry cyclicality, which is one of the semiconductor industry's distinguishing features.[31]
    c. A further aspect of the industry's steep learning curves is the need for the research and at least some of the production facilities to be in close geographic proximity. This permits the many adjustments required to improve performance and yields and to adapt new equipment, production processes, and design features while adjusting to changing market conditions. The learning and synergies among university research, private laboratories, production and changing customer needs is a recognized feature of the semiconductor industry.[32]
    d. This does not mean that the benefits of proximity require all production to be located within a particular geographic area. This would imply a freezing of the allocation of global semiconductor production that would be neither possible nor desirable. On the other hand, having no on-shore production would inevitably erode the quality and robustness of research, design, equipment and materials production in the United States.

---

[29]Despite industry cyclicality, the semiconductor industry achieved a 16.1 percent compound annual growth rate (CAGR) from 1975 to 2000. Growth during this period was driven by technological advances, the increasing pervasiveness of electronics in society, and the increasing capability of the semiconductors that powered new products and systems. This growth rate began to slow gradually starting in the mid-1980s, reaching about 15 percent in 1998. The severity of the 2001 downturn then prompted a reevaluation of the industry's long-term growth rate. With semiconductor sales of $213 billion in 2004, the rate is now expected to be in the 8-10 percent range. The Semiconductor Industry Association forecast, released in June 2005, reflects this consensus and predicts a CAGR for the industry of 9.2 percent from 2004 to 2008. Accessed on the Semiconductor Industry Association Web site at *<http://www.sia-online.org/iss_economy.cfm>*.

[30]European Semiconductor Industry Association, *The European Semiconductor Industry: 2005 Competitiveness Report,* op. cit.

[31]Kenneth Flamm, "Factors Underpinning Cyclicality in the Semiconductor Industry," in National Research Council, *Productivity and Cyclicality in Semiconductors: Trends, Implications, and Questions*, op. cit., pp. 61-64.

[32]See, for example, European Semiconductor Industry Association, *The European Semiconductor Industry: 2005 Competitiveness Report,* op. cit.

e. The rapid rate of innovation means that products embedding semiconductor devices often have short life cycles. At the same time, the rate of price-performance improvement in the semiconductor industry is very rapid. Consequently, changes in the semiconductor market can occur very quickly, and established markets—and market leaders—can be swiftly displaced. In order to adjust constantly to this rapid pace of change, the semiconductor industry needs to be highly flexible and able to rapidly adopt new designs and new technologies. In this way, the rapid rate of innovation translates into high-capital requirements.[33]

f. The high capital costs for fabrication facilities and the high R&D intensity make the industry sensitive to incentive schemes to attract and retain foreign and domestic semiconductor investment. Major competitors such as China, Japan, Korea, Malaysia, Singapore, Taiwan, and Europe have developed such incentive schemes. In the United States, these have been generated primarily at the state level (e.g., New York and Texas). It is important to recognize that lower wages and lower social costs are not determining factors in locational choices for semiconductor investments, whereas the existence of favorable incentive schemes, in particular tax regimes, is often the main source of competitive advantage.[34]

Recommendations

1. ***Data and Modeling Challenges:*** Policies to foster continued improvement in the nation's productivity and growth can be best developed with better data on prices and better models for prediction.
   a. Serious gaps in data prevent a full accounting of semiconductor-related prices.
   b. Although information technology is altering product markets and business organizations, a fully satisfactory model of the semiconductor industry remains to be developed. Such a model would derive the demand for semiconductors from investments in information technology in response to rapidly falling IT prices. An important objective is to determine the product cycle for successive generations of new semiconductors endogenously.

[33]European Semiconductor Industry Association, *The European Semiconductor Industry: 2005 Competitiveness Report,* op. cit., p. 2.

[34]European Semiconductor Industry Association, *The European Semiconductor Industry: 2005 Competitiveness Report,* op. cit., p. 4. This perspective is not new. As Laura Tyson famously observed, "the semiconductor industry has never been free of the invisible hand of government intervention." Laura Tyson, *Who's Bashing Whom? Trade Conflict in High Technology Industries*, Washington, D.C.: Institute for International Economics, 1992, p. 85.

2. ***Investments in Research and Training:*** Substantial investments in research and a well-trained workforce are needed if we as a nation are to continue to benefit from the growth and technological development offered by a vibrant, internationally competitive semiconductor industry.
   a. ***National Investments:*** National investments are necessary to provide the research and development facilities as well as a trained workforce well grounded in the disciplines—especially physics, chemistry, mathematics, computer science, and engineering—that underpin research and manufacturing in the semiconductor, computer component, and software industries.[35] Such investments are also necessary for the nation to realize its substantial investments in nanotechnology, given its potential impact on computing, telecommunications, and semiconductor technology. National investments in these IT-related disciplines also continue to be important to capitalize on potential advances in biotechnology and related biomedical research. Scientific advances are increasingly multidisciplinary efforts and information technologies are often a key element in such advances.[36]
   b. ***Acquiring and Retaining Talent:*** Continual progress is also necessary in visa processing in order to attract and retain qualified foreign engineers and scientists. This includes increases in the number of H-1B visas; automatic visa extensions for international students who receive advanced degrees in science, technology, engineering, mathematics, and other fields of national need from U.S. institutions; as well as more permanent opportunities for science and engineering graduates to remain and contribute to the United States economy.[37] Compensation packages

[35]Previous analysis by the Board on Science, Technology, and Economic Policy (STEP) of trends in federal research funding found that "there has been a significant reduction in federal funding in certain of the physical science and engineering fields. These include fields whose earlier advances contributed to the surge in productivity and economic growth of the late 1990s and fields that underlie progress in energy production and conservation, pollution abatement, medical diagnosis and treatment, and other national priorities." National Research Council, *Trends in Federal Support of Research and Graduate Education,* Stephen A. Merrill, ed., Washington, D.C.: National Academy Press, 2001.

[36]This point was highlighted in the recent study of public-private partnerships led by Gordon Moore. See National Research Council, *Capitalizing on New Needs and New Opportunities: Government-Industry Partnerships in Biotechnology and Information Technologies*, Charles W. Wessner, ed., Washington, D.C.: National Academy Press, 2001.

[37]The need for these actions is emphasized in the recent Congressionally mandated NAS/NAE/IOM study, *Rising Above the Gathering Storm: Energizing and Employing America for a Brighter Economic Future,* op. cit. See, in particular, Actions C-4 through C-6, which call for continuing improvements in visa processing for international students, providing a one-year automatic visa extension to international students who receive doctorates or the equivalent in fields of national need, and the institution of a new skills-based, preferential immigration option that would significantly raise the chances of an applicant with doctoral-level education in science and engineering. Problems with H-1B visas remain acute, acting as an impediment to the retention and recruitment of high-value

for technology workers are also a factor for remaining competitive in attracting and retaining qualified scientists and engineers.[38]

3. ***Partnering for Innovation:*** To sustain the technology trajectory envisaged by Moore's Law requires advanced research to overcome emerging technological "brick walls" that threaten continued rapid advance. Substantial public funding and cooperative partnerships in semiconductor research are necessary if we are to continue to reap the benefits of remaining on the trajectory set out by Moore's Law and for the United States to remain a robust global center for the research, development, and production of semiconductors.
   a. Sustained research and development is necessary for the semiconductor industry to overcome the limits of CMOS (complementary metal-oxide semiconductor) and develop post-CMOS technologies. Initiatives in this regard include fostering research and developments in nanotechnology and molecular electronics to replace and/or extend the life of advanced CMOS manufacturing technologies.[39]
   b. To maintain the innovative pace of the industry, with the attendant benefits for the U.S. economy, national investments in university research programs that explore and develop promising technologies are needed.
   c. Additional government investments in university research for programs that support and move promising technologies closer to commercialization are increasingly important to maintain the innovative pace of the semiconductor industry.[40]
   d. Public-private partnerships, involving cooperative research and development activities among industry, universities, and government laboratories can play an instrumental role in accelerating the development

---

human capital. See *Wall Street Journal*, "Lopsided Immigration Policy Could Induce Brain Drain," June 22, 2005, p. A17.

[38]National Research Council, *Building a Workforce for the Information Economy*, Washington, D.C.: National Academy Press, 2001, pp. 69-79.

[39]The Committee on the Future of Supercomputing of the National Research Council similarly discusses the imperative to innovate the next generation of hardware for supercomputers. See National Research Council, *Getting Up to Speed: The Future of Supercomputing*, Susan L. Graham, Marc Snir, and Cynthia A. Patterson, eds., Washington, D.C.: The National Academies Press, 2005, Chapter 5.

[40]Previous analysis by the STEP Board has underscored the importance of innovation partnerships such as The Small Business Innovation Research (SBIR) Program and the Advanced Technology Program in contributing to the development of new technologies. See National Research Council, *The Advanced Technology Program: Assessing Outcomes*, Charles W. Wessner, ed., Washington, D.C.: National Academy Press, 2001, p. 39. See also National Research Council, *The Small Business Innovation Research Program: An Assessment of the Department of Defense Fast Track Initiative,* Charles W. Wessner, ed., Washington, D.C.: National Academy Press, 2000.

of new technologies and products.[41] Industry experts believe that such partnerships provide the most promising strategy for sustaining Moore's Law, given that the semiconductor industry's ability to make smaller, faster, and cheaper integrated circuits is limited by the growing inability of individual firms to pay for the increasingly expensive research needed to achieve needed innovations.[42]

e. In addition to pre-competitive research partnerships at the horizontal level (e.g., among semiconductor device manufacturers), vertical partnerships focused on integrated capacities along the supply chain are seen as increasingly important. The objective of the vertical partnerships is to ensure competitiveness across the development and production chain through synergistic relations among suppliers, manufacturers, and users of semiconductors.[43]

f. Finally, it is important to recognize that innovation partnerships are increasingly international efforts, even as global markets for high-technology industries are increasingly competitive. Governments can serve as a facilitating agent to create the necessary credibility, commitment, and mutual trust among private firms in the formation of research consortia.[44] International research consortia, such as SEMATECH and IMEC, demonstrate the benefits of such global cooperation by reducing the risks and costs associated with the development of new semiconductor technologies and the standards for their application.

4. Wider adoption of road-mapping exercises by the computer and computer component industries (along the lines of the ITRS conducted by the semiconductor industry) can contribute to the industries' ability to remain on the

---

[41]Public-private partnerships involving cooperative research and development among industry, government, and universities can play an instrumental role in introducing key technologies to the market. For an overview of the conditions necessary for successful partnerships (including industry leadership, development and use of technology roadmaps, shared costs, and regular assessment) see National Research Council, *Government-Industry Partnerships for the Development of New Technologies: Summary Report*, op. cit., pp. 13-16.

[42]A recent study by SEMI estimates that research required for continued scaling of integrated circuit devices, even without another wafer size increase, will cost some $16.2 billion by 2010. However, the equipment and materials suppliers, to whom the burden of research has shifted from chipmakers, are predicted to be able to afford an annual R&D budget of $10.4 billion, creating a $6 billion gap. SEMI, "Semiconductor Equipment and Materials: Funding the Future," October 2005. Accessed at *<http://content.semi.org/cms/groups/public/documents/homepervasive/p036611.pdf>*.

See also Phil LoPiccolo, "The Six Billion Dollar Gap," *Solid State Technology*, February 2006; and Robert Haavind, "Chipmaking's Tough Economic Road Ahead," *Solid State Technology*, March 2006.

[43]European Semiconductor Industry Association, *The European Semiconductor Industry: 2005 Competitiveness Report: Executive Summary,* op. cit., p. 51.

[44]For an assessment of the limits and challenges of international cooperation, see HWWA, IfW, and NRC, *Conflict and Cooperation in National Competition for High-Technology Industry,* Washington, D.C.: National Academy Press, 1996, pp. 54-61.

growth path predicted by Moore's Law with its contribution to the pace of innovation and growth.[45]

C. MAINTAINING U.S. TECHNOLOGY LEADERSHIP IN SEMICONDUCTORS

Findings

1. The semiconductor industry is a key driver for the future of advanced technologies in the United States. The distinctive features of this industry enable it to foster new opportunities for economic growth and support the global competitiveness of the U.S. products and services.
    a. As a ***technology enabler***, the semiconductor permits the invention and use of a variety of valuable applications. Semiconductor-based information technology is also a ***general purpose technology***, shared across a wide variety of uses. There is also the possibility of substantial ***network effects*** that can amplify the impact of advances in semiconductor technology.[46] Indeed, semiconductor technologies already underpin a variety of products ranging from personal computers and mobile phones, to solutions and services, especially those provided through the Internet. Taken together, these features of information technology mean that advances in semiconductor technology can have substantial impacts on long-run economic growth.
    b. Through their pervasiveness, semiconductors have become keys to the competitiveness of the products of a broad range of new and "traditional" industries. For example, the automobile industry uses semiconductor-based information technologies to design and manufacture vehicles at lower cost. In addition, onboard microprocessors increasingly monitor fuel use and driver safety. By improving quality, lowering costs, creating new features, and increasing customization, microelectronics can help differentiate the products of traditional industries, helping make U.S. firms more globally competitive.[47]

---

[45]An example is the roadmap exercise by the U.S. Display Consortium, which develops platform technologies for flat-panel displays. See <*http://www.usdc.org/*> for additional information.

[46]On the role of general purpose technologies, see Timothy Bresnahan and Manuel Trajtenberg, "General Purpose Technologies: Engines of Growth?" *Journal of Econometrics*, 65(1):83-108, 1995. See also Elhanan Helpman, "General Purpose Technologies and Economic Growth: Introduction" in *General Purpose Technologies and Economic Growth,* Elhanan Helpman, ed., Cambridge, MA: The MIT Press, pp. 1-13, 1998.

[47]Similarly, for the role information technology is playing in reviving the competitiveness of the U.S. textile and apparel industry, see Lenda Jo Anderson et al., "Discovering the Process of Mass Customization: A Paradigm Shift for Competitive Manufacturing," National Textile Center Annual Report, 1995.

c. The unique contributions of this industry are reflected in the policies and programs of almost all major global participants. The perceived importance of the industry to long-term economic growth and technological competency has resulted in an impressive array of policies, partnerships, subsidies, and investments that are intended to create, nurture, and retain the design, development, production, and refinement of semiconductors and related technologies.[48]

2. Even as some production moves offshore, semiconductors remain important for the U.S. economy. The production and use of semiconductors are major contributors to the growth and dynamism of the U.S. economy. Access to and use of advanced semiconductors contribute to many national missions, not least national security.[49]
   a. Ever faster and cheaper semiconductors are recognized as key components in sustaining the productivity growth that the U.S. economy has experienced since 1995. The end of a two-decade slowdown in U.S. productivity growth that took hold in the 1970s and that coincided with a significant erosion of the country's industrial power can be traced to a sudden speed-up in the rate of decline of semiconductor and computer prices.[50]
   b. The semiconductor industry is U.S. manufacturing's star performer. In 2003, the industry saw worldwide sales of $166 billion, of which $80 billion were sales in the United States. (The U.S. semiconductor industry invests $14 billion in R&D, an amount that represents 17 percent of sales, with another 14 percent of sales [$10 billion] going towards the acquisition of capital equipment.) The industry provides about 226,000 jobs in the United States. While relatively small compared to the total U.S. workforce, these jobs are well-paid and have a disproportionately positive impact on the U.S. economy because of their remarkable productivity levels.[51]
   c. In light of the unique contributions of semiconductors to national growth and technical capacity in this information age, fostering a vigorous semiconductor industry in the United States is important for the nation's

---

[48]These multiple programs and incentives are documented by Tom Howell, "Competing Programs: Government Support for Microelectronics," in National Research Council, *Securing the Future: Regional and National Programs to Support the Semiconductor Industry*, op. cit.

[49]National Research Council, *Productivity and Cyclicality in Semiconductors: Trends, Implications, and Questions,* op. cit., 2004. See also National Research Council, *Securing the Future: Regional and National Programs to Support the Semiconductor Industry*, op. cit.

[50]Dale W. Jorgenson, Mun S. Ho, and Kevin J. Stiroh, *Productivity, Volume 3: Information Technology and the American Growth Resurgence,* op. cit., 2005, Chapter 9.

[51]Ibid. See also Semiconductor Industry Association, Industry Facts and Figures. Accessed at <*http://www/sia-online.org/ind_facts.cfm*> on December 13, 2005.

long-term security and for its contributions to economic growth, productivity, and technological know-how.

3. ***Structural Change:*** The structure of the semiconductor industry is changing, creating new vulnerabilities and challenges for U.S. leadership in this strategically important industry, while at the same time creating new opportunities.
   a. The semiconductor industry is characterized by a very high intensity of R&D and high levels of capital expenditures in semiconductor fabrication facilities. A modern fabrication facility is now in the $2 billion to $3 billion range, increasingly out of the reach of many manufacturers. Reflecting these high forced capital costs, the industry is highly sensitive to both the global research infrastructure and the incentives and disincentives that condition the financial returns on its investments.[52]
   b. The foundry model is now a significant component of the industry, having begun with Taiwanese government encouragement and investment in 1987.[53] Foundry-based companies tend to spend a smaller percentage of their sales on R&D than do traditional integrated device manufacturers.[54] Consequently, one potential impact of the growth in the foundry model maybe a fall-off in industry-sponsored research into process engineering, possibly posing long-term challenges to the pace of innovation in semiconductor manufacturing techniques. A shortfall in R&D directed to process improvements is unlikely to be made up by semiconductor design firms, which are smaller with less commitment to process technologies.
   c. At the same time, the foundry model permits easier entry and greater competition through the development and commercial application of

[52]The European Semiconductor Industry Association (ESIA) estimates a 220 percent higher return on a facility in East Asia as a result of revenue incentives. ESIA, "The European Semiconductor Industry 2005 Competitiveness Report." Accessed at <*http://www.eeca.org/pdf/final_comp_report.pdf*>. Locational competition continues to accelerate. In 2005, Germany attracted an AMD fab with very substantial incentives, including $700 million in loan guarantees, $500 million in grants and allowances, and $320 million in equity contributions. Recently, the State of New York provided about $1 billion in incentives for AMD to locate near Albany a new 300-mm wafer plant. It will make chips based on a 32-nm process and will take about $3.2 billion in capital to build. See *PC Magazine*, "AMD to Build Factory in New York," June 26, 2006.

[53]According to Wikipedia.org, in microelectronics, a foundry refers to a factory where devices such as integrated circuits are manufactured. The foundry model describes how businesses separate the design process from the manufacturing of these microdevices. For an account of the modern evolution of the foundry model, see Jon Sigurdson, "VSLI Revisited—Revival in Japan," Working Paper No. 191, Tokyo: Institute of Innovation Research of Hitotsubashi University, p. 50, April 2004.

[54]See comments by George Scalise on "The Foundry Phenomenon," National Research Council, *Productivity and Cyclicality in Semiconductors: Trends, Implications, and Questions,* op. cit., p. 15.

semiconductors with new cost and performance features that would otherwise not be available as rapidly or as attractively priced.[55]

d. While globalization offers many benefits, if design, materials and equipment, and manufacturing capabilities move outside the United States, the United States risks losing the critical mass necessary for its leadership and autonomy in semiconductor technologies and equipment.[56]

4. The movement of the semiconductor industry offshore is not uniquely the result of market forces. Semiconductors are produced and traded in a globally integrated market, and firms have significant interest in locating facilities in rapidly growing markets. Yet at the same time, national policies often condition international competition in semiconductors. Indeed, the policies of other nations and regions may well pose challenges to U.S. leadership in this sector.[57]
   a. Publicly supported, location-based competition for high-value-added, high-growth industries is one of the hallmarks of the global economy. Many governments in East Asia and Europe have adopted comprehensive and effective policies to attract, create, and retain semiconductor firms and related industries within their national economies.[58] Governments adopt and finance these policies in order to secure national capacity and autonomy in this enabling technology, as well as the increased competitiveness and future government revenue associated with the semiconductor industry.[59]
   b. The United States has no current comparable national effort to retain and maintain the industry. SEMATECH, the government-industry partnership, was founded in 1987 at the height of the Japanese industry's challenge to U.S. producers, and it proved effective in improving U.S. manufacturing capabilities.[60] SEMATECH's major international efforts

---

[55]Jack Harding, "Current Trends and Implications: An Industry View," in National Research Council, *Software, Growth, and the Future of the U.S. Economy*, Dale W. Jorgenson and Charles W. Wessner, eds., Washington, D.C.: The National Academies Press, 2006.

[56]President's Council of Advisors on Science and Technology, *Sustaining the Nation's Innovation Ecosystems*, Washington, D.C.: Executive Office of the President, 2004, pp. 9 and 14.

[57]National Research Council, *Securing the Future: Regional and National Programs to Support the Semiconductor Industry*, op. cit.

[58]The European semiconductor industry affirms that national incentives are shaping international competition and calls for its governments and the EU technology programs to become more focused. See ESIA, "The European Semiconductor Industry 2005 Competitiveness Report," op. cit.

[59]For a review of national and regional programs, see Thomas Howell, "Competing Programs: Government Support for Microelectronics," in National Research Council, *Securing the Future: Regional and National Programs to Support the Semiconductor Industry*, op. cit., pp. 254-284.

[60]See National Research Council, *Securing the Future: Regional and National Programs to Support the Semiconductor Industry*, op. cit., pp. 41-43. See also Kenneth Flamm, "SEMATECH Revisited: Assessing Consortium Impacts on Semiconductor Industry R&D" in National Research Council,

began in 1996, and since 1998 SEMATECH has been a fully international research consortium. International SEMATECH receives no federal funding, though it has recently benefited from significant state support from Texas and more recently for its expansion in New York State.[61] It now faces competition from IMEC, located in Flanders, a novel and effective research consortium supported with European Union, national government, and Flanders regional funds as well as contributions from private companies from all major regions.[62]

c. Also fundamental to the development of the globally competitive semiconductor industry has been the opening of global markets that are relatively free from unfair trade practices. U.S.-led trade initiatives to open foreign semiconductor markets played a key role in establishing the conditions for international competition based on price and quality among multiple vendors.[63]

Recommendations

1. To better address the technical challenges faced by the semiconductor industry and to better ensure the foundation for continued progress, more attention to the conditions and policies shaping locational decisions for this enabling industry is warranted.
2. Renewed attention to encouraging and retaining a capable high-tech workforce is necessary.[64] Most importantly, additional resources for university-

---

Securing the Future: Regional and National Programs to Support the Semiconductor Industry, op. cit., pp. 254-281.

[61]*Electronic News*, "SEMATECH Adds 4 International Members," June 21, 1999.

[62]Jon Sirgurdson suggests that the more independent and flexible IMEC approach to employing new technical opportunities may reflect an advantage in international competition for research dollars and expertise. Sirgudson, "VSLI Revisited—Revival in Japan," op. cit., pp. 48-49. IMEC receives some 35 million euros in "core support" from the Flanders regional funds.

[63]In the mid-1980s, the United States, Japan, and Canada entered into agreements to eliminate tariffs, first on semiconductors, and then on parts of computers. Separately, the Japanese government and industry agreed to a market opening initiative and foreswore dumping in the United States and other markets. This had several effects: A Korean DRAM industry grew up under this regime, and the United States stayed in memory chip production, not only in DRAMs (*Dynamic Random Access Memory,* a type of memory used in most personal computers) but also in EPROMs (*Erasable Programmable Read-Only Memory*) and was able to enter the flash memory market. By the mid-1990s, the Japanese market had become completely open to foreign semiconductor makers. Vigorous price competition among a multiple vendor base ensured that prices and costs would decline sharply, enabling the growth of IT use and, with it, the Internet and globalization phenomena. For a detailed analysis of trade in semiconductors, see Kenneth Flamm, *Mismanaged Trade*, op. cit.

[64]George Scalise, "Industry Perspective on Semiconductors," in National Research Council, *Securing the Future: Regional and National Programs to Support the Semiconductor Industry*, op. cit., pp. 35-42. Google's Wayne Rosing reiterated this need at the Committee's workshop on "Software, Growth, and the Future of the U.S. Economy," National Research Council, *Software, Growth, and the*

based research in related disciplines, such as physics, chemistry, materials sciences, and engineering, are required.[65]

3. The federal and state governments should undertake measures that strengthen the attractiveness of the United States as a location for semiconductor research and production.
4. Three-way partnerships among industry, academia, and government are needed to catalyze progress in the high-cost area of future process and design. These partnerships would:
    a. Sponsor more initiatives that encourage collaboration between universities and industry, especially through student training programs, in order to generate research interest in solutions to impending and current industry problems.
    b. Increase funding for successful current programs. For example, the Focus Center Research Program developed by the Semiconductor Research Corporation could usefully be augmented through increased direct government funding.[66] These centers also represent opportunities for collaborative research with other federal outreach programs, such as those supported by the National Science Foundation.
    c. Create incentives for students. Augmented federal support for programs that encourage research in semiconductors would attract professors and graduate students. In addition, specific incentive programs could be established to attract and retain talented graduate students.[67]

5. Active, rapid, and effective enforcement of international trade rules through the World Trade Organization (WTO) and other forums is needed to maintain

---

Future of the U.S. Economy, op. cit. See also remarks by Craig Barrett at a Semiconductor Industry Association event commemorating the 40th anniversary of Moore's Law. Access news release at <*http://www.sia-online.org/pre_release.cfm?ID=355*>.

[65]See the corresponding recommendation in National Research Council, *Securing the Future: Regional and National Programs to Support the Semiconductor Industry*, op. cit., pp. 88-89. See also more recent related recommendations concerning strengthening the nation's traditional commitment to science and engineering in NAS/NAE/IOM, *Rising Above the Gathering Storm: Energizing and Employing America for a Brighter Economic Future*, op. cit.

[66]The Microelectronics Advanced Research Corporation (MARCO) funds and operates university-based research centers in microelectronics technology. Its charter initiative, the Focus Center Research Program (FCRP), is designed to expand pre-competitive, cooperative, long-range applied microelectronics research at U.S. universities. Each Focus Center targets research in a particular area of expertise. In addition to strengthening ties between industry and the university research community, this model concentrates resources on the areas of microelectronics research that are critical in maintaining industry growth. More information can be accessed at <*http://fcrp.src.org/Default.asp?bhcp=1*>.

[67]NAS/NAE/IOM, *Rising Above the Gathering Storm: Energizing and Employing America for a Brighter Economic Future*, op. cit.

a competitive global market in semiconductors.[68] The successful U.S. case against the Chinese value-added tax (VAT) rebate is a case in point.[69]

D. DECONSTRUCTING THE COMPUTER

Findings

1. Computers are widely recognized to be important components of economic growth and improved productivity associated with the New Economy, with the upward shift in economic growth coincident with declines in the prices of computers and related equipment.
   a. Indeed, it seems that the upward shift in the rate of economic growth in the mid-1990s coincided with a sudden, substantial, and rapid decline in the prices of computers (from 15 percent annually to about 28 percent annually after 1995, per the data graphed in Figure 2[70]) accompanied by significant increases in computing power and function.
   b. This upward shift in growth also coincided with a shift in the rate of decline in price for memory and logic devices (from 40 percent annually to about 60 percent annually after 1995 per the data graphed in Figure 1).[71]
   c. Indeed, recent estimates suggest that between 40 and 60 percent of the decline in computer prices in the late 1990s, and perhaps 20 to 30 percent of the declines in communications equipment and consumer

[68]WTO and other trade rules do not include a mandate for the enforcement of competition/antitrust policy; they only permit it. Enforcement of competition policy is necessary. Recently, Samsung Electronics of South Korea has agreed to plead guilty to charges of participating in an international conspiracy to fix prices in the DRAM market to the tune of $300 million, settling with the U.S. Department of Justice Antitrust Division. See *Electronic News*, "Samsung Faces $300M DoJ Fine for Price Fixing," October 13, 2005.

[69]In rapidly evolving industries, such as semiconductors, it is important that remedial trade actions to correct merchantalistic policies (e.g., through discriminatory taxation) be both prompt and effective. For example, the United States and China agreed in 2004 on a resolution to their dispute at the World Trade Organization (WTO) regarding China's tax refund policy for integrated circuits. U.S. exports of integrated circuits to China were subject to a 17 percent value-added tax (VAT). However, China taxed domestic products significantly less, allowing firms producing integrated circuits in China to obtain a partial refund of the 17 percent VAT, lowering the effective VAT rate on domestic products to as low as 3 percent in some cases. These measures contributed to a very significant competitive advantage and a powerful incentive for inward investment, reflected in the surge in new production facilities in China.

[70]The output price index referred to in Figure 2 is the GDP deflator, but differs from the typical BEA GDP deflator due to methodology. We impute a capital service flow for government and consumer durable capital and use Tornqvist aggregation to add components of GDP.

[71]Dale W. Jorgenson and Kevin J. Stiroh, "Raising the Speed Limit: U.S. Economic Growth in the Information Age," in National Research Council, *Measuring and Sustaining the New Economy*, op. cit.

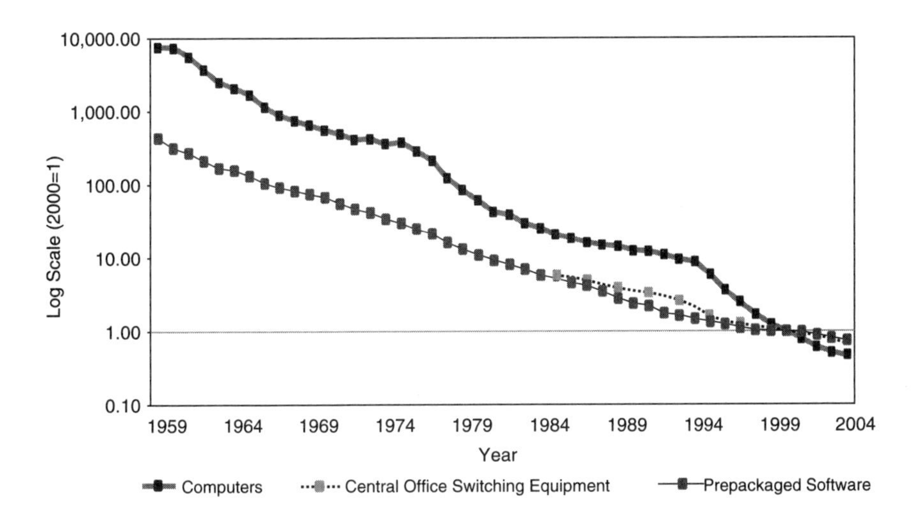

FIGURE 2 Relative prices of computers, communications equipment, and software, 1959-2004.
NOTE: All price indexes are divided by the output price index.

electronics, are directly attributable to declines in price for the semiconductor inputs used in these products.[72]

2. The Moore's Law phenomenon of "faster, cheaper, better" semiconductors is also present in the computer and computer component industries, accelerating the pace of technological innovation and lowering costs to consumers.

3. Increased computing power enables a wider and more complex range of applications for both the specialist as well as the general user.
   a. The increased computing power in new and improved computer microprocessors is readily apparent to those "power users" who use their computers for scientific computing or modeling, games, and video and audio processing.
   b. It also of advantage to less demanding household computer users, improving the quality of their interface to their computers and enhancing their ability to handle high-resolution audio, video, and image files that are increasingly available over the Internet. Increased computing power also improves the speed and responsiveness of household computer

[72]See Ana Aizcorbe, Kenneth Flamm, and Anjum Kurshid, "The Role of Semiconductor Inputs in IT Hardware Price Declines," in *Hard to Measure Goods and Services: Essays in Honor of Zvi Griliches*, E. Berndt, ed., Chicago, IL: National Bureau of Economic Research, forthcoming.

users in executing traditional applications (like word processing and spreadsheets), their ability to use features available in these traditional applications, and their ability to run multiple other applications (like anti-virus checkers, firewalls, multimedia plug-ins, messaging software, and search engines) concurrently with traditional applications without significantly degrading performance.

4. Economists require accurate measures of the performance of computers and computer components in order to understand their contributions to economic growth.
   a. Developing a useful measure of computer performance through time is a challenge because the nature of the computer is changing in many ways. For example, some experts forecast that the emergence of grid computing and Web-based services will change information technology from assets that firms own—in the form of computers, software, and other devices—to a service they purchase from utility providers.[73]
   b. Hedonic price indexes provide a proven method for adjusting for quality differences in computers across time. Using this method requires improved performance measures for computers and computer components.[74]

5. Microprocessors are the single largest semiconductor expenditure in computers, and there are some signs that quality-adjusted price declines for this input have recently slowed significantly. This slowdown appears to be linked to a recently struck "brick wall" that now limits continuing improvements in microprocessor operating speeds.[75] If this continues, it suggests the pos-

[73]See Mike Nelson, "Moving Computing to the Grid," in National Research Council, *The Telecommunications Challenge: Changing Technologies and Evolving Policies,* Charles W. Wessner, ed., Washington, D.C.: The National Academies Press, 2006. Also, see Nicholas Carr, "The End of Corporate Computing," *MIT Sloan Management Review*, 46(3):67-73, 2005.

[74]Gregory Chow pioneered the use of hedonic techniques for constructing a constant quality index of computer prices in research conducted at IBM. See Gregory C. Chow, "Technological Change and the Demand for Computers," *American Economic Review*, 57(5):117-130, December 1967. In 1985, BEA incorporated constant quality price indexes for computers and peripheral equipment constructed by IBM into the National Income and Product Accounts (NIPA). The economic interpretation of these indexes by Jack Triplett brought the rapid decline of computer prices to the attention of a very broad audience. See Jack Triplett, "The Economic Interpretation of Hedonic Methods," *Survey of Current Business*, 66(1):36-40, January 1986. Triplett has also provided exhaustive surveys of research on hedonic price indexes for computers. See, for example, Jack Triplett, *Handbook on Hedonic Indexes and Quality Adjustments in Price Indexes: Special Application to Information Technology Products*, Paris: Organisation for Economic Co-operation and Development, 2004.

[75]Improvements in microprocessor clock rates have historically been the largest single contributing factor to declines in prices for these components. For evidence on these points see Kenneth Flamm, "Economics of Innovation in the Microprocessor Industry," Working Paper, University of Texas at Austin, February 2006. Flamm's analysis is confirmed by recent Bureau of Labor Statistics (BLS)

sibility that we may soon see a lessening of the pace of price declines for computers, which ultimately will have negative economic implications for the wider economy as a whole.

Recommendations

1. The computer component industry has developed a variety of formal and informal measures to gauge the relative performance of its products. Further development of these measures and subsequent incorporation into the National Income and Product Accounts should enable improved analysis and policies to sustain the contributions of computers and computer components to economic growth.

2. Wider adoption of technology roadmaps and enhanced government-industry cooperation can improve our capacity to reinforce the growth path predicted by Moore's Law in the computer and computer component industries.

3. Given the apparent slowdown in microprocessor speeds, accelerated research investment in methods and tools to lessen the cost of writing software that can utilize multiple computer processor cores "in parallel" would seem to be a priority, particularly given the potentially large economic payoffs in maintaining the pace of technological advance in IT.[76]

E. SOFTWARE'S CRUCIAL ROLE

Findings

1. Software is the means by which we interface with the information and communications technologies that underpin the modern economy.
   a. The United States economy is highly dependent on software, with businesses, public utilities, and consumers among those integrated within complex software systems.
   b. Computer systems, such as those used by businesses, integrate software and hardware with detailed knowledge about the context of the application. A workforce that understands both the nature of business processes

data showing that there has been a substantial slowdown in the rate of decline of prices for microprocessors. This is NAICS 334413 (Semiconductor and Related Device Manufacturing) and can be found in the BLS detailed report on the Producer Price Index (PPI). Accessed at <*http://www.bls.gov/ppi/ppidr_t01-09.pdf*>.

[76]A recent National Academies study by the Committee on the Future of Supercomputing identifies this problem. See National Research Council, *Getting Up to Speed: The Future of Supercomputing*, op. cit.

as well as information technology is necessary to develop and maintain such systems.

2. The structure of software is highly complex. While better software permits operations at unparalleled levels of sophistication, this complexity also creates significant economic and national security vulnerabilities.
   a. As software has become more complex, safeguarding it has become more difficult. Attacks against software—in the form of both network intrusions and infection attempts—have also grown substantially in recent times. Moreover, the economic impact of such attacks is increasingly significant.
   b. Thus, a major challenge lies in creating software code that is relatively error free, virus-resistant, robust against change, and capable of scaling reliably to incredibly high volumes while ensuring that it can integrate seamlessly and reliably to many other software systems in real time.

3. Tracking software prices and aggregate investments in software, and hence their impact on the economy, is a challenge given the unique and embedded nature of software.[77]
   a. Software is complex in structure, and the market for software is different from that of other goods and services. Software can be easily duplicated (often at low cost) and the service life of software is often hard to anticipate. The nature and functions of software also evolve over time, requiring the development of quality-adjusted price indexes for various types of software.
   b. In addition, most of the nation's software capability is embedded in firms that are not classified as the software industry. These sectors include financial services, health care informatics, telecommunications, defense, aerospace, and automobiles. Tracking the value contribution of software in these sectors is challenging because the capability is thoroughly integrated into the organization, and despite this, that capability depends closely on technologies developed in the information technology sectors, including software, computers, and IT services.
   c. The Bureau of Economic Analysis (BEA) distinguishes among three types of software—prepackaged, custom, and own-account software. Prepackaged software is sold or licensed in standardized form and is delivered in packages or electronic files downloaded from the Internet.

[77]For a detailed overview of the role of IT in the national accounts and methods used by BEA (and BLS) to construct appropriate prices, see Bruce Grimm, Brent R. Moulton, and David B. Wasshausen, "Information-Processing Equipment and Software in the National Accounts," in *Measuring Capital in the New Economy,* Carol Corrado, John Haltiwanger, and Daniel Sichel, eds., Chicago, IL: University of Chicago Press, 2005, pp. 363-402.

Custom software is tailored to the specific application of the user and is delivered along with analysis, design, and programming services required for customization. Own-account software consists of software created for a specific application. However, at present, only price indexes for prepackaged software hold performance constant.[78]

d. Given that only the prices of prepackaged software are adequately represented in the official system of price statistics, software prices are in a statistical blind spot. A major challenge lies in constructing constant quality price indexes for custom and own-account software.

4. Open-source software is promising. Open-source software development has proven to be a significant and successful way of creating software that is more robust.[79]
   a. The policy challenge lies in fostering incentives for individuals to develop basic software components through open-source coordination, while ensuring that once they are built, they will be widely available at low cost so that future development can be stimulated.[80]
   b. Measuring open-source software in the national accounts is a challenge given the distribution in prices and value-added services and the need to separate business services from software.

5. Software suffers from relatively slow measured productivity growth. In contrast to trends found in other information technology sectors, productivity growth in software development does not appear to be as significant.
   a. There is limited progress in automating the production of software. Software writing—particularly at the creative or high end—remains in many respects a cottage industry.
   b. Complementarities among skill sets needed to develop software mean that scaling up and speeding up production are difficult.

---

[78]Robert P. Parker and Bruce T. Grimm, "Recognition of Business and Government Expenditures on Software as Investment: Methodology and Quantitative Impacts, 1959-1998," Washington, D.C.: Bureau of Economic Analysis, November 2000.

[79]For example, Apple Inc. notes that "using Open Source methodology makes Mac OS X a more robust, secure operating system, as its core components have been subjected to the crucible of peer review for decades. Any problems found with this software can be immediately identified and fixed by Apple and the Open Source community." See <*http://www.apple.com/opensource/*>. For an empirical analysis showing fewer and more rapid bug fixes with open-source software, see Jennifer Kuan, "Open Source Software as Lead User's Make or Buy Decision: A Study of Open and Closed Source Quality," Palo Alto, CA: Stanford Institute for Economic Policy Research, 2002.

[80]See comments by Hal Varian in National Research Council, *Software, Growth, and the Future of the U.S. Economy*, op. cit.

6. The software workforce is highly differentiated and includes computer scientists and engineers of varying caliber. Increasingly, this labor pool is dispersed around the world.
   a. There are a relatively small number of high-quality software developers. Industry participants suggest that there are only a very limited number of such "superstars"—those with productivity that is 20 to 100 times better than that of average software developers—and that this talent is scattered worldwide.[81]
   b. Competition for high-quality software developers is accelerating. The international competition for this limited pool of highly skilled labor is acute and is expected to accelerate.[82]

7. Software and hardware play interdependent roles in enhancing the productivity of information technology. More widespread use of multiple-core processors, needed to overcome the apparent slowdown of processor clock rates, will require additional software development.
   a. The apparent slowdown in improvement of processor clock rates means that much continuing price-performance improvement in computer hardware will increasingly require harnessing the power of microprocessors with multiple cores. Making effective use of such multi-core processors is likely to be less difficult for business computers (like servers) that serve applications to many different users and can make easy use of this greater processor power to serve more users.
   b. However, harnessing multiple processor cores on a single more powerful application is much more complex, and requires writing or reengineering software to split it up into multiple threads that can operate in parallel.

Recommendations

1. Software price indexes, especially for own-account and custom software, must be upgraded to hold software performance constant. Without adjustment for quality, these indexes present a distorted picture of software prices as well as software output and investment.
   a. Advances in developing software price indexes, including current work by BEA on function points, hedonic techniques, and other methodologies, should be supported.[83] These advances can improve statistical

---

[81]William Raduchel, "The Economics of Software," in National Research Council, *Software, Growth, and the Future of the U.S. Economy,* op. cit.

[82]Wayne Rosing, "Hiring Software Talent," in National Research Council, *Software, Growth, and the Future of the U.S. Economy,* op. cit.

[83]BEA has recently launched a software pricing project for custom and own-account software using function points. Work by Q/P Management Group and the Analysis Group is expected to produce new price indexes for custom and own-account software for the U.S. national accounts. A function

information on firm investments in customized software applications such as own-account and custom software.[84]

b. Adoption of common standards across the Organisation for Economic Co-operation and Development (OECD) and beyond should also be further encouraged. Wider use of standards can improve our knowledge about investments in software in what is a global industry and facilitate the tracking of software outsourcing.[85]

c. Active and informed participation by standards organizations such as the National Institute of Standards and Technology (NIST), in close consultation with industry associations, are necessary if the United States is to participate effectively in the process.

2. The United States needs to foster the expert workforce needed to develop and maintain the computer systems so critical to the nation's economy and security.
   a. Developing the basis for a better-trained workforce begins with strengthening K-12 education. At the secondary level and beyond, scholarships are needed to attract more U.S. students, including women and minorities, to pursue training in computer science and related fields.[86]
   b. More adaptive immigration policies are also required to attract and retain in the United States foreign students who have been trained in American universities—especially those who are exceptionally talented.[87] In addition, visa restrictions that prevent or impede highly talented software developers from working in the United States should be revised.
   c. The nature of the market for software superstars is poorly understood.[88] Given the increasingly apparent "bottleneck" role of software development in limiting the continued growth in the New Economy, a major research effort aimed at understanding the economics of the software industry and software labor markets would seem highly desirable.

---

point metric is a means of measuring software size and productivity. It uses functional, logical entities such as inputs, outputs, and inquiries that tend to relate more closely to the functions performed by the software. See John J. Marciniak, ed., *Encyclopedia of Software Engineering,* New York, NY: John Wiley & Sons, 1994, pp. 518-524.

[84]David Wasshausen, "A BEA Perspective: Private Fixed Software Investment," in National Research Council, *Software, Growth, and the Future of the U.S. Economy,* op. cit.

[85]See comments by Dirk Pilat, "What is in the OECD Accounts and How Good is it?" in National Research Council, *Software, Growth, and the Future of the U.S. Economy,* op. cit.

[86]NAS/NAE/IOM, *Rising Above the Gathering Storm: Energizing and Employing America for a Brighter Economic Future*, op. cit., Chapter 5.

[87]National Research Council, *Policy Implications of International Graduate Students and Postdoctoral Scholars in the United States*, Washington, D.C.: The National Academies Press, 2005.

[88]William Raduchel, "The Economics of Software," in National Research Council, *Software, Growth, and the Future of the U.S. Economy*, op. cit.

3. Sustaining the productivity gains from information technology includes maintaining existing computer systems and, given the limited lifespan of such systems, investing in the development of future systems.
   a. Computer systems have a limited lifespan. Improved metrics are necessary if firms are to properly capitalize their software expenses, anticipate the liabilities at the end of the software system's life cycle, and plan for future systems.
   b. Improvements in custom coding, software-oriented architecture, and Web-based services will be necessary to sustain the productivity gains from software.
   c. Interdisciplinary training that combines computer science with business management, finance, and other application fields is necessary to develop the expertise required to build and maintain the information technology systems of the future.

4. The slowdown in the improvement of processor clock rate, which historically has been the focus for applications running on supercomputers, will soon become vastly more important on desktop computers containing new, multicore microprocessors. Significant national investment in basic research on software development methodologies and tools for "parallelizing" applications to make use of multiple processors would seem to be a very worthwhile endeavor, with potentially significant economic impacts in maintaining the pace of the New Economy.[89]

F. THE TELECOMMUNICATIONS CHALLENGE

Findings

1. Communications technology is crucial for the rapid development and diffusion of the Internet, perhaps the most striking manifestation of information technology in the American economy. By storing, sorting, and distributing vast information very quickly and at very low cost over communications networks, the Internet may be potentially very important in the longer run for the continued growth in output and improved productivity of the United States and other knowledge economies.[90] Communications equipment is also

[89]As mentioned earlier, this problem was recently noted and a congruent recommendation made by the National Academies Committee on the Future of Supercomputing. See National Research Council, *Getting Up to Speed: The Future of Supercomputing*, op. cit. See also Kenneth Flamm, "The Coming IT Slowdown: Technological Roots and Economic Implications," Working Paper, LBJ School of Public Policy, University of Texas, October 2005.

[90]The rapid growth of the Internet (and particularly the World Wide Web) is captured by Metcalfe's Law, which observes that the power of the network increases exponentially by the number of computers connected to it. This is because computers added to the network not only use it as a resource

an important market for semiconductors. Switching and terminal equipment rely heavily on semiconductor technology, so that product development in communications often reflects improvements in semiconductors.

2. Advances in telecommunications equipment, however, are derived from a variety of sources. Technological advance in fiber optics, microwave broadcasting, and communications satellites, as well as for switches and routers that are used to send and receive data, have progressed at rates that in some cases outrun even the dramatic pace of semiconductor development.[91] The convergence of these advanced technologies is a powerful source of innovation.
   a. For example, phenomenal progress has been made possible in part by advances in Dense Wavelength Division Multiplexing (DWDM), a technology that sends multiple signals over an optical fiber simultaneously. Installation of DWDM equipment, beginning in 1997, has doubled the transmission capacity of fiber optic cables every 6 to 12 months.
   b. Prices for communications gear have also been estimated to have fallen by about 8 to 10 percent per year between 1994 and 2000, although this is about half as fast as the decline in the price for computers.[92]

3. The United States recently witnessed significant investments in telecommunications equipment, corresponding with the dot-com boom of the late 1990s. Much of this communications investment was in the form of the transmission gear, connecting data, voice, and video terminals to switching equipment.
   a. Investments in communications equipment in the United States are on par with those for computers. Over the course of the 1990s and continuing into the present decade, expenditure on communications has been around $100 billion per year, representing a little over 10 percent of total equipment investment in the United States.

---

but also potentially contribute resources available to other users, increasing the value and choices of the network.

[91]The rate of improvement in the capacity of communications technology was first suggested by George Gilder, who observed that the maximum transmission rate for telecoms is tripling every year. This means that if computer power doubles every 18 months (per Moore's Law), then communications power doubles every 6 months. See also estimates of 85 percent per annum growth since 1983 by Debra J. Aron, Ken Dunmore, and Frank Pampush, "Worldwide Wait? How the Telecom Act's Unbundling Requirements Slow the Development of the Network Infrastructure," *Industrial & Corporate Change*, 7(4):615-621, 1998.

[92]Mark Doms, "The Record to Date: Quality Adjusted Prices for Equipment," in National Research Council, *The Telecommunications Challenge: Changing Technologies and Evolving Policies*, op. cit.

b. At the same time, there have been large swings in the U.S. investment in communications, with investment in communications gear falling 35 percent during the recession of the early 2000s.

4. Although massive investments in the nation's high-capacity Internet backbone have created excess capacity in long-haul facilities, a variety of factors—regulation among them—have slowed the build-out of the crucial last mile.
   a. By creating highly technology-specific industry rules, and by attempting to promote competition by requiring incumbents to share the local loops of their network with rivals, the Telecommunications Act of 1996 may have, according to some experts, inadvertently inhibited investment needed to diffuse high-bandwidth access over the last mile.[93]
   b. While broadband adoption has grown quickly in recent years, demand for broadband adoption appears to be slowing. A recent survey by the Pew organization shows that 32 percent of the adult population in the United States does not use the Internet, a number that held steady for the first 6 months of 2005.[94]
   c. According to the International Telecommunications Union, the United States significantly lags other advanced nations in high-speed broadband access, ranking sixteenth in the world in broadband penetration in 2005.[95] However, these data relate to diffusion to households, whereas information about diffusion to businesses (which has important consequences for productivity) is limited.
   d. Our limited knowledge about the scope of broadband diffusion and adoption inhibits policies needed to better capitalize on the nation's

[93]Robert Litan and Roger G. Noll, "The Uncertain Future of the Telecommunications Industry," Brookings Working Paper, December 3, 2003, Washington, D.C.: The Brookings Institution. Interpretations vary on the impact of the 1996 Telecommunications Act. Some experts believe that competition for the provision of broadband was already taking place in most major downtown areas in many of the largest cities of the United States before the Telecommunications Act. See Glenn Woroch, "Local Network Competition," in *Handbook of Telecommunications Economics*, Martin Cave, Sumit Majumdar, and Ingo Vogelsang, eds., New York, NY: Elsevier, 2002. Others believe that the Act did not deter the build-out of the nation's cable network. For example, see Jonathan E. Nuechterlein and Philip J. Weiser, *Digital Crossroads: American Telecommunications Policy in the Internet Age*, Cambridge, MA: The MIT Press, 2005.

[94]John B. Horrigan, "Broadband Adoption at Home in the United States: Growing but Slowing," Pew Internet and American Life Project, paper presented to the 33rd Telecommunications Policy Research Conference, September 25, 2005.

[95]The International Telecommunication Union (ITU) reports that in 2005 the five top nations for broadband network market penetration were Korea, Hong Kong, the Netherlands, Denmark, and Canada. The ITU ranked the United States sixteenth in broadband penetration. ITU Strategy and Policy Unit Newslog, August 8, 2005. Accessed at <*http://www.itu.int/osg/spu/newslog/CategoryView,category,Broadband.aspx*>.

substantial investments in information technology and infrastructure, limiting the potential for sustained growth in the economy.

5. Wireless broadband can help overcome some of the limitations associated with traditional wired broadband access—not least the costs associated with wiring the last mile.
   a. While wireless broadband has been in limited use to date due to relatively high subscriber costs and technological challenges such as those related to obstacle penetration, rapid advances in technology are likely to overcome such challenges.
   b. Some industry experts believe that the emerging WiMAX standard promises to resolve a number of problems that confront existing wireless protocols such as WiFi.[96]

6. The convergence of data, voice, video, wireless, and public and private networks into an end-to-end infrastructure, now under way, is challenging business models and regulatory frameworks alike. The commoditization of information is ushering a major shift from distinct vertically integrated industries that are focused on particular products or services towards a more horizontal platform that supports the movement of content and application services moving across IT networks.[97]
   a. This convergence is changing the terms of competition across industries. While there was once a major separation between the telecom and cable industries, for example, these businesses are likely to overlap and offer similar kinds of services.
   b. Regulators face new challenges as telecommunications services are increasingly becoming blended, with voice, data, and video transmitted in commoditized packets over the air or through a wire. The end of "stovepiping" also poses new challenges for consumers.[98] Consumers confronted with a proliferation of Internet services, operating systems, and devices may look for service that is integrated and easy to use.[99]
   c. Replacing "silo" or sector-specific communications regulation with a policy framework that address the emerging horizontal technology plat-

[96]David Lippke, "The Wireless Wildcard," in National Research Council, *The Telecommunications Challenge: Changing Technologies and Evolving Policies*, op. cit.

[97]William Raduchel, "The End of Stovepiping," in National Research Council, *The Telecommunications Challenge: Changing Technologies and Evolving Policies,* op. cit.

[98]Stovepiping refers to the retrieval of information from unconnected databases—in particular the situation that exists when it is necessary to "climb out" of one database in order to climb down into another.

[99]Lisa Hook, "Serving Consumers on Broadband," in National Research Council, *The Telecommunications Challenge: Changing Technologies and Evolving Policies,* op. cit.

forms is needed to complement the networked characteristic of the New Economy.[100]

7. The value of intellectual property is increasingly recognized, and the use of patents has increased dramatically. Yet, businesses have faced serious challenges of protecting intellectual property in the era of digital distribution. Recently available technologies, for example, have allowed consumers to share music and content with each other for free.[101] The trend towards improved security of intellectual property, such as the recent success of the iPod and related legitimate forms of on-line music diffusion, is encouraging and can help stimulate the creation of new content and applications.[102]

8. The move towards virtualization, grid computing, and Web services is leading to a major shift in the nature of information technology assets from computers, software, and myriad related components that companies own to services that firms purchase from on-line utility providers.[103]

Recommendations

1. The varying complexity and rates of technical innovation make the contribution of telecommunications equipment to productivity growth a challenge to measure. Current BEA methodologies for making intertemporal comparisons in price and quality understate true price declines in communications equipment because they do not fully track evolving technological changes.[104]
   a. For most of the 1990s, manufacturers focused on features such as greater port density, faster speeds, and support for an increasing number of communication protocols in designing new switches and routers. After the 2001 collapse in demand for telecommunications equipment, manufacturers began to differentiate their products in new and innovative ways that are difficult to quantify on a quality-adjusted basis. Better data and analysis are needed to get a clear idea of what happened with regard to

[100]Peter Tenhula, "Technological Change and Economic Opportunity: The View from the Federal Communications Commission," in National Research Council, *The Telecommunications Challenge: Changing Technologies and Evolving Policies,* op. cit.

[101]However, in the case of the music-swapping service Napster, a U.S. federal appeals court ruled in 2001 that the file-sharing Internet company must stop trading in copyrighted material.

[102]Steve Metalitz, "The View from the Copyright Industries," in National Research Council, *The Telecommunications Challenge: Changing Technologies and Evolving Policies,* op. cit.

[103]Nicholas G. Carr, "The End of Corporate Computing," op. cit.

[104]BEA estimated that prices for communications gear fell an average of 3.2 percent per year between 1994 and 2000. Recent analysis by Marc Doms however suggests that communications equipment prices actually fell about 8 to 10 percent over that period. Mark Doms, "Communications Equipment: What has Happened to Prices?" Federal Reserve Bank of San Francisco Working Paper, 2003-15.

technological change in and prices of communications equipment from 2001 on.

   b. Valuing the improvements built into new switches and routers is difficult. While the Producer Price Index has tried to address some of these changes using hedonic techniques, data that consistently identify important current period product characteristics and transaction prices are not yet readily available.[105] Research into alternative quality valuation techniques and improved data transparency is required to respond to the technological changes in telecommunications equipment. BEA and other statistical agencies require increased funding to follow evolving trends in the communications arena with more accuracy.

2. As noted above with regard to software, greater attention to standards and the national and international process of their establishment is required. The economic stakes of standard setting are of great consequence. Some nations and regions see standards as a competitive tool and devote substantial resources to this end. The role and resources of the National Institute of Standards and Technology have to be seen in this light. The standard-making process must be recognized as a key component of U.S. competitiveness and provided commensurate resources and policy attention.
   a. Uncertainty created by a multiplicity of standards and a lack of clarity in regulatory policy are retarding progress in the growth of wireless and fiber networks needed to convey this commoditized information to the curb.[106]
   b. Technical standards, especially for wireless devices, are an important element in sustaining U.S. success in the global economy. Without effective standard-making capabilities and active U.S. participation in international standard-making bodies, the United States will not be able to maximize its advantages.[107]

3. The supply as well as the demand side of the market for high-speed Internet access needs to be elaborated. While international comparisons show that U.S. broadband adoption for households lags that of other countries, relatively little is known about factors that affect the broadband adoption path

---

[105]For additional perspective on the types of technological changes in telecom equipment that, at least conceptually, could be valued in a hedonic model, see Michael Holdway, "Confronting the Challenge of Estimating Constant Quality Price Indexes for Telecommunications Equipment in the Producer Price Index," Bureau of Economic Analysis Working Paper, 2002.

[106]See Action D-4 on ensuring ubiquitous broadband Internet access in NAS/NAE/IOM, *Rising Above the Gathering Storm: Energizing and Employing America for a Brighter Economic Future*, op. cit.

[107]See National Research Council, *Standards, Conformity Assessment, and Trade into the 21st Century*, Washington, D.C.: National Academy Press, 1995.

in the United States, particularly for businesses. Further data are required to understand the scope and nature of broadband use by businesses, and more study is required to understand why a significant percentage of households are not linked to the computer and Internet culture that is central to the new, more productive U.S. economy.[108]

4. Revising outdated regulation and addressing issues of security and intellectual property protection are necessary for the nation to realize productivity gains from advances in communications technology.
   a. To address the technology convergence now under way, sector-specific telecom regulation—for radio, CDs and DVDs, television, telephony, and mobile telephony—must be replaced with a more horizontal (as opposed to a vertically stovepiped) regulatory regime.
   b. Internet-based services, through such means as grid computing, have the potential to improve productivity. Improved security in the broadband network, which stands as a major obstacle to more widespread use of the technology, has to be addressed.
   c. Developing a legitimate market for copyrighted materials over broadband—for entertainment, software, services, video games, research and reference works—is needed for the long-term viability of these industries.

G. GLOBALIZATION AND OFFSHORING

Findings

1. Rapid progress in information and communications technologies combined with continuing efforts to liberalize international trade and investment in services, have increased the tradability of services and created new types of tradable services.[109] This has led to a new wave of globalization in the services sector, with offshoring of particular types of services now becoming increasingly common.

2. The development of the Internet, in particular, has generated great economic opportunity, facilitated growth, and improved peoples' lives in many ways. The Internet is also making possible new forms of business organization.

[108] John B. Horrigan, "Broadband Adoption at Home in the United States: Growing but Slowing," op. cit.

[109] Desirée van Welsum and Xavier Reif, "Potential Offshoring: Evidence from Selected OECD Countries," OECD-DSTI-ICCP, July 2005.

a. The declining cost of computing and communications is changing the structure of industrial organizations—replacing vertically integrated industries with newly economical horizontal platforms.[110]
b. A new model for value delivery to businesses is based on an intelligent infrastructure, where this infrastructure spans the globe. Here the network becomes a repository of intelligence across a broad spectrum of applications, such as caching, security, multicasting, and network management.[111]

3. Greater horizontal integration across the globe created by faster and cheaper information and communications technologies is opening new opportunities for U.S. firms to compete worldwide.
   a. The globalization of IT hardware and software research, production, and trade, as well as new competition from lower-wage producers, have helped reduce IT prices and should contribute to technological advance, helping to maintain Moore's Law.[112] These, in turn, have facilitated the wider adoption of information technologies, making available the efficiencies that arise from their use.[113]
   b. From the perspective of many firms, cost savings through the outsourcing of research and production offshore provides a compelling business motive. It has rapidly become "best practice" for the business plans of new high-technology manufacturing and service companies.[114]

4. Globalization is also giving rise to a shift in the comparative advantage of nations, raising major policy challenges—including questions concerning national competency, capacity, and autonomy in strategic technologies.[115]
   a. Access to low-cost, highly trained workers located around the world and the advantages of round-the-clock development cycles—made possible by low-cost computing and communications—have led to the relocation abroad of many business functions that can be outsourced. This phenom-

[110]See comments by William Raduchel, "The End of Stovepiping," in National Research Council, *The Telecommunications Challenge: Changing Technologies and Evolving Policies*, op. cit.

[111]David S. Isenberg, "Rise of the Stupid Network," *Computer Telephony*, pp. 16-26, August 1997.

[112]24/7 research cycles and larger R&D teams made possible by less expensive skilled labor can accelerate research and product cycles. See *Businessweek*, "The Rise of India," December 8, 2005.

[113]Catherine Mann, *High-tech and the Globalization of America*, op. cit.

[114]See comments by Jack Harding, "Current Trends and Implications: An Industry View," in National Research Council, *Software, Growth, and the Future of the U.S. Economy*, op. cit. See also recent comments by Michael Mortiz, a prominent venture capitalist, who notes that "we can barely imagine investing in a company without at least asking what their plans are for India." See *Businessweek*, "The Rise of India," op. cit.

[115]See the recent findings by the National Academies' Committee on Materials Research and Development in National Research Council, *Globalization of Materials R&D: Time for a National Strategy*, Washington, D.C.: The National Academies Press, 2005.

enon is popularly known as "offshoring." In addition to the moving of existing jobs to other countries, another dimension of offshoring is likely to be the creation of new service jobs in countries other than the United States—a trend that will be difficult to measure.

b. By diffusing technology and research capabilities around the world, globalization enables other countries, including newly emerging economies like China and India, to pursue technological leadership in key areas. Some analysts and business leaders believe that this diffusion of expertise abroad is eroding U.S. comparative advantages in high-technology innovation and manufacturing.[116]
c. For the United States, the economic and strategic risks associated with offshoring include a loss of within-country expertise and future talent, dependency on other countries for key technologies, and increased vulnerability to political and financial instabilities abroad.[117]
d. The Internet enables trade in services to a greater extent than before. The types of jobs subject to offshoring are increasingly moving from low-end services (such as call centers and help desks) to higher-technology services (such as software and microchip design, business consulting, medical analysis, and drug development) where the United States has traditionally enjoyed a comparative advantage.
e. These trends notwithstanding, some analysts believe that there are limits to near-term globalization. On the demand side, recent experience seems to reveal that offshoring is proving successful for businesses only in

[116]Some recent studies have questioned whether "offshoring" is simply another form of trade with mutual benefits. David Levy, for example, argues that reducing wages through offshore outsourcing leads to wealth creation for shareholders but not necessarily for countries and employees, and that many displaced workers have difficulty "trading up" to higher-skilled jobs. The result, he noted, is the creation of global commodity markets for particular skills and a shift in the balance of market power among firms, workers, and countries. David L. Levy, "The New Global Political Economy," *Journal of Management Studies*, 42(3):685, May 2005.

This caution has been echoed in industry. Andy Grove of Intel has noted that firms need to strike a balance between maximizing shareholder value and their obligation to U.S. workers who helped build the nation's technology industry but who are now being replaced by cheaper labor. *Forbes,* "Grove Says U.S. Is Losing Its Edge in High-Tech Sector," October 10, 2003. See also the discussion by William Bonvillian, "Offshoring Policy Options," in National Research Council, *Software, Growth, and the Future of the U.S. Economy*, op. cit.

[117]In his dissent from the mainstream economic consensus about outsourcing and globalization, Paul Samuelson has argued that the assumption that the laws of economics dictate that the U.S. economy will benefit in the long run from all forms of trade, including the outsourcing of call-center and software programming jobs abroad, is a "popular polemical untruth." Trade does not always work to all parties' advantage, according to Samuelson. See Paul Samuelson, "Where Ricardo and Mill Rebut and Confirm Arguments of Mainstream Economists Supporting Globalization," *Journal of Economic Perspectives*, 18(3), 2004.

selected contexts due to operational and management limitations.[118] On the supply side, some believe that there are only a limited number of low-wage knowledge workers abroad in the market today who possess the necessary language skills, technical qualifications, and related abilities needed for successful international collaboration.[119] Even if these limits to near-term globalization hold true, it is not clear whether this condition will persist. Further research is necessary to estimate how quickly these limitations may be overcome.

Recommendations

1. Although the offshoring phenomenon—particularly the offshoring of service-sector jobs—is a topic of much currency, the scope of the phenomenon remains to be adequately documented. Despite extensive media attention, there is relatively little hard information about the causes and impact of offshoring on manufacturing and service-sector employment in the United States or on other related economic and structural developments.[120]
   a. A sustained effort to measure the dimensions and implications of offshoring is necessary for informed policymaking. Further research is needed to make adequate evaluations of the effects of outsourcing, including the impact of high-tech job creation abroad rather than in the United States.[121] To overcome the lack of appropriate, adequate data for

[118]According to the McKinsey Global Institute, internal barriers within firms, most notably operational issues, management attitude to offshoring, and structural issues can limit demand. Adding that external regulatory barriers also play a small role overall, it concludes that the "potential for global resourcing varies depending on the industry." See McKinsey Global Institute, *The Emerging Global Labour Market*, Part I: "Demand for Offshore Talent and Sector Cases," 2005.

[119]McKinsey Global Institute, *The Emerging Global Labour Market,* Part II: "Synthesis of Findings: Supply of Offshore Talent," 2005. Some statements about the numbers and qualifications of Indian and Chinese engineers may be overstated, while U.S. graduates and qualifications may be understated. See Gary Gereffi and Vivek Wadhwa, "Framing the Engineering and Outsourcing Debate: Placing the United States on a Level Playing Field with China and India," Duke University School of Engineering, December 12, 2005.

[120]While preliminary analysis has not found that outsourcing of business services has had much effect on the growth of the U.S. economy (Ralph Kozlow and Maria Borga, "Offshoring and the US Balance of Payments," Washington, D.C.: Bureau of Economic Analysis, 2004), popular evaluations argue that the world of technology is becoming increasingly international (see, for example, Thomas Friedman, *The World is Flat*, New York, NY: Farrar Straus & Giroux, 2005).

[121]The 2006 report of the Association for Computing Machinery's Job Migration Task Force similarly finds that "while offshoring will increase, determining the specifics of this increase is difficult given the current quantity, quality, and objectivity of data available. Skepticism is warranted regarding claims about the number of jobs to be offshored and the projected growth of software industries in developing countries." Association of Computing Machinery, "Globalization and Offshoring of Software," William Aspray, Frank Mayadas, and Moshe Y. Vardi, eds., New York, NY, 2006. Lastly, assessments of jobs outsourced do not necessarily take into account the impact of foreign investment, increasingly for countries such as India and China, on U.S. employment.

this analysis, the necessary resources should be made available to provide better information both to policymakers and to the general public about the evolution and performance of the American economy.

b. In any event, we must recognize that the global competitive environment is shifting. According to several recent reports, the pace of global competition is accelerating and the United States will need to redouble support for existing strengths (e.g., research), strengthen proven commercialization programs, and experiment with new efforts.[122]

2. To thrive in the globally competitive environment, the United States has to maintain its technological leadership. This requires continuing investments in the nation's science and technology infrastructure. This, in turn, requires both substantial investments in science and technology education as well as experimentation with policy mechanisms that can capitalize on these investments.
   a. The United States' current leadership in high technology draws from substantial federal investments starting in the postwar period in the nation's science and technology infrastructure. Key elements of this policy have included building a system of research universities and attracting foreign talent though scholarships and by providing academic freedoms and research facilities not available elsewhere. This institutional capital has to be upgraded and adapted to new needs and opportunities if the United States is to maintain its leadership as a knowledge-based economy.
   b. U.S. information technology firms, and the U.S. economy, will forgo the benefits of leadership unless they can attract the best human resources, garner sufficient research funding, develop and support robust mechanisms for technology transfer,[123] and maintain an internationally competitive business environment.
   c. Four specific steps to retain and develop the semiconductor industry and U.S. research capacity in this sector would include:
      i. Substantial additional research funding should be provided at the 10 percent per year recommended in the NAS/NAE/IOM report,

[122]For example, see Council on Competitiveness, *Innovate America: Thriving in a World of Challenge and Change,* Washington, D.C., 2004.

[123]In this regard, the recent flattening of research funding and the elimination of funding for new awards from the Advanced Technology Program are troubling. The recent NAS/NAE/IOM report, *Rising Above the Gathering Storm: Energizing and Employing America for a Brighter Economic Future*, op. cit., argues for additional R&D funding and new efforts to transition them to market. Previous Academy analysis, led by Intel's Gordon Moore, documented the substantial positive accomplishments of the Advanced Technology Program. See National Research Council, *The Advanced Technology Program: Assessing Outcomes*, op. cit., pp. 87-98. The NRC study found the program's industry-driven, cost-shared approach to funding promising technological opportunities to be effective. It also found a high standard for assessment. Indeed, the quality of the assessment lends credence to the program's evaluation of its accomplishments.

Rising Above the Gathering Storm,[124] with particular emphasis on the physical sciences and engineering.

ii. U.S. policymakers would benefit from evaluations of current worldwide policy and programs in support of this enabling industry.[125]

iii. The federal government should continue to provide support for proven mechanisms to transfer technology and commercialize promising technologies.[126]

iv. The United States needs to adopt measures to encourage research in the United States, including providing an attractive business environment involving competitive tax regimes, infrastructure support, cooperative research programs, and generous tax credits for corporate R&D. These measures are essential to maintain a competitive advantage for the U.S. economy in the locational competition for high-technology industries.[127]

H. NEW ARCHITECTURE FOR THE U.S. NATIONAL ACCOUNTS

Findings

1. The U.S. national accounts were originally constructed to deal with issues arising from the Great Depression of the 1930s, and the basic architecture of the national accounts has not been substantially altered in 50 years. In the meantime, the success of monetary and fiscal policies has shifted the policy focus from stabilization of the economy to enhancing the economy's growth potential. In addition, the economy is confronted with new challenges arising from rapid changes in technology and globalization.

2. America's economy is large and diverse, and it is not surprising that accounting for this vast range of economic activity requires a decentralized statistical system. The major agencies involved in generating the national accounts include the Bureau of Economic Analysis (BEA) in the Department of Commerce, the Bureau of Labor Statistics (BLS) in the Department of Labor, and the Board of Governors of the Federal Reserve System.

---

[124]NAS/NAE/IOM, *Rising Above the Gathering Storm: Energizing and Employing America for a Brighter Economic Future*, op. cit.

[125]The Board on Science, Technology, and Economic Policy has an analysis of foreign programs under way that is focused on best practices, *Comparative Innovation Policy: Best Practice in National Technology Programs*, under the direction of William Spencer.

[126]National Research Council, *The Advanced Technology Program: Assessing Outcomes*, op. cit.

[127]Similar priorities have been identified in Europe. See ESIA, "The European Semiconductor Industry 2005 Competitiveness Report," op. cit.

3. The BEA has responsibility for the core system of accounts, the National Income and Product Accounts. The BLS generates employment statistics, wage and salary data, and productivity statistics, as well as almost all of the underlying price information. The Board of Governors produces the Flow of Funds Accounts, including balance sheets for major financial sectors. Many other agencies and private sector organizations provide source data for the national accounts.
   a. As an illustration, both BEA and BLS measure industry output. BEA's estimates are used to allocate the GDP to individual industries. BLS generates its own estimates in arriving at measures of industry-level productivity growth. Unfortunately, the BEA and BLS estimates of industry output do not always agree.
   b. As a second illustration, the Board of Governors generates a measure of national saving from the income statements and balance sheets that comprise the Flow of Funds Accounts. BEA produces an estimate of national saving from the income and product accounts. Although both estimates agree that the saving rate has declined sharply over the past 20 years, they employ different data sources and sometimes arrive at conflicting results.

Recommendations

1. The U.S. national accounts require a new architecture to guide the future development of this decentralized system. The National Income and Product Accounts, the productivity statistics, and the Flow of Funds have different origins, reflecting diverse objectives and data sources. However, they are intimately linked. An important motivation for developing a new architecture is to integrate the different components and make them as consistent as possible.[128]
2. An important objective of the new architecture is to combine the data sources employed by BEA and BLS in order to arrive at a common set of estimates. This is a crucial ingredient in long-term projections of the U.S. economy that depend on the disparate trends in productivity in key industries, such as information technology producers and intensive users of information technology.

3. Another goal of this new architecture is to bring the flow of funds and the national income accounts into consistency in order to provide better data for

[128]More details about the new architecture are presented in Dale W. Jorgenson, J. Steven Landefeld, and William Nordhaus, eds., *A New Architecture for the U.S. National Accounts,* Chicago, IL: University of Chicago Press, 2006. Accessed at <*http://www.nber.org/books/CRIW-naccts/index.html*>.

anticipating future financing needs of both public and private sectors.[129] This proposed system of accounts integrates the National Income and Product Accounts with the productivity statistics generated by BLS and balance sheets produced by the Federal Reserve Board. The proposed system would feature GDP, as does the National Income and Product Accounts; however, GDP and domestic income are generated along with productivity growth. BEA's accounts for reproducible assets and the U.S. International Investment Position are extended to encompass a balance sheet for the U.S. economy as a whole.

4. The cost of capital for productive assets employed in the U.S. economy provides a unifying methodology for integrating the National Income and Product Accounts generated by BEA and the productivity statistics constructed by BLS.[130] The next step is to develop a complete version of the BLS productivity statistics that is consistent with a new system of official industry accounts recently released by BEA.[131]

5. To further explore these proposals for a new architecture, additional resources should be made available. The drivers of the U.S. economy have evolved, indeed shifted quite dramatically, and it is essential that a new architecture for the national accounts be put into place to better capture this new reality.[132]

---

[129]The key elements of the new architecture are outlined in a "Blueprint for Expanded and Integrated U.S. Accounts," by Dale W. Jorgenson and J. Steven Landefeld, 1995. Accessed at <*http://post.economics.harvard.edu/faculty/jorgenson/papers/Blueprint_051905.pdf*>.

[130]A detailed set of productivity statistics for the United States is presented by Dale W. Jorgenson, Mun S. Ho, and Kevin J. Stiroh, *Productivity, Volume 3: Information Technology and the American Growth Resurgence,* op. cit.

[131]Access at <*http://www.bea.doc.gov/bea/newsrel/gdpindnewsrelease.htm*>.

[132]The BEA is currently working with the National Science Foundation on accounts for research and development that could ultimately lead to recognition of R&D investment and capital stocks. The results of this accounting should assist future evaluations of high-tech investment and its effects on the economy.

III

SUMMARY OF THE NRC CONFERENCES ON THE NEW ECONOMY

The NRC Conferences on the New Economy

Faster, better, and cheaper semiconductors and computers as well as software and telecommunications equipment have led, especially over the past decade, to the widespread adoption and use of modern information and communications technologies. This, in turn, is rapidly ushering fundamental changes to the way in which (and the rapidity with which) goods and services are developed, manufactured, and distributed around the world and the way in which individuals and businesses everywhere consume, interact and transact. This "New Economy" poses new challenges, requiring new approaches to economic measurement and policy analysis.

To this end, the National Academies' Board on Science, Technology, and Economic Policy (STEP) has since 2000 held a series of workshops to better understand the New Economy phenomenon and to develop policies needed to sustain the positive contribution of modern information and communications technologies to U.S. growth and competitiveness. This section of the report summarizes and provides background for some of the key issues raised over the course of the five conferences hosted by the STEP Board (listed in the Preface) on Measuring and Sustaining the New Economy.

The proceedings of each of these conferences have been published in separate volumes by The National Academies Press. Although the technologies of the industries considered at these conferences continue to evolve rapidly, the reports nonetheless capture conceptual issues of continued policy relevance to the industry leaders, academics, policy analysts, and others who participated in these workshops.

MOORE'S LAW AND THE NEW ECONOMY

At the time of the STEP Board's first conference in 2000, many economists were still reluctant to proclaim a technology-driven New Economy if only because there were few or no data reflecting economy-wide returns to the substantial investments made by U.S. business in new information and communications technologies.[1] Throughout the 1970s and 1980s, Americans and American businesses regularly invested in ever more powerful and cheaper computers and communications equipment. They assumed that advances in information technology—by making more information available faster and cheaper—would yield higher productivity and lead to better business decisions.

The expected benefits of these investments did not appear to materialize—at least in ways that were being measured. Even in the first half of the 1990s, productivity remained at historically low rates, as it had since 1973. This phenomenon was called "the computer paradox," after Robert Solow's casual but often repeated remark in 1987: "We see the computer age everywhere except in the productivity statistics."[2]

Raising the Speed Limit

At the National Academies first conference on the New Economy, however, Dale Jorgenson pointed to new data that showed that the U.S. economy was undergoing a fundamental change.[3] While growth rates had not returned to those of the "golden age" of the U.S. economy in the 1960s, he noted, new data did reveal an acceleration of growth accompanying a transformation of economic activity. This shift in the rate of growth by the mid-1990s, he added, coincided with a sudden, substantial, and rapid decline in the quality-adjusted prices of semiconductors from an average of 15 percent annually before 1995 to 28 percent annually after 1995.[4]

In response to the rise in capability of computers and drop in price, investment in semiconductor-based technologies exploded, leading to a positive impact on economic growth. Jorgenson and Stiroh have calculated that computers' con-

[1]For the views of a notable skeptic, see Robert J. Gordon, "Does the 'New Economy' Measure Up to the Great Inventions of the Past?" *Journal of Economic Perspectives*, American Economic Association, 14(4):49-74, 2000.

[2]R. Solow, "We'd Better Watch Out," *New York Times Book Review*, July 12, 1987. The implications of the Solow Productivity Paradox have since been actively discussed. For example, see J.E. Triplett, "The Solow Productivity Paradox: What Do Computers Do to Productivity?" *Canadian Journal of Economics*, 32(2):309-34, April 1999.

[3]National Research Council, *Measuring and Sustaining the New Economy*, Dale W. Jorgenson and Charles W. Wessner, eds., Washington, D.C.: National Academy Press, 2002.

[4]Dale W. Jorgenson and Kevin J. Stiroh, "Raising the Speed Limit: U.S. Economic Growth in the Information Age," in National Research Council, *Measuring and Sustaining the New Economy*, op. cit., 2002, Appendix A.

tribution to growth rose more than five-fold, to 0.46 percent per year in the late 1990s. Software and communications equipment contributed an additional 0.30 percent per year for 1995-1998. And their preliminary estimates through 1999 revealed further increases for all three categories.[5] Jorgenson thus made the case for "raising the speed limit"—that is, for revising upward the intermediate-term projections of growth for the U.S. economy.[6]

The Role of Moore's Law

Moore's Law describes the speed at which semiconductor technology develops. Semiconductors are the core enablers for the wide array of information and communications technology. The pace of semiconductor development is, therefore, critical to the development of the broader range of computing and telecommunications technologies that are the basis for modern economic processes.

Moore's Law is based on a prediction made by Gordon Moore in a 1965 paper titled "Cramming More Components onto Integrated Circuits," where he noted:

> The complexity for minimum component costs has increased at a rate of roughly a factor of two per year. Certainly, over the short term, the rate of increase is a bit more uncertain, although there is no reason to believe it will not remain nearly constant for at least 10 years. That means by 1975, the number of components per integrated circuit for minimum cost will be 65,000.[7]

Extrapolating this trend (see Figure 1), Gordon Moore predicted an exponential growth of chip capacity at 35 to 45 percent per year through 1975.[8]

Gordon Moore revised his original prediction in 1975 (the endpoint of his earlier projection) stating that increases in components per chip would continue, approximately doubling every 2 years, rather than every year.[9] Believing that human ingenuity would further sustain the growth of chip capacity, he noted that manufacturers were using "finer scale microstructures" to engineer higher density of components per chip.

As Kenneth Flamm pointed out at the National Academies' 2001 conference on semiconductors, the idea popularly known today as "Moore's Law" (drawn from but not identical to Gordon Moore's predictions) anticipates the doubling of

---

[5]Ibid.

[6]Ibid.

[7]See Gordon E. Moore, "Cramming More Components onto Integrated Circuits," *Electronics*, 38(8), April 1965.

[8]For a historical overview of Moore's Law, see Kenneth Flamm, "Moore's Law and the Economics of Semiconductor Price Trends," in National Research Council, *Productivity and Cyclicality in Semiconductors: Trends, Implications, and Questions*, Dale Jorgenson and Charles Wessner, eds., Washington, D.C.: The National Academies Press, 2004.

[9]See Gordon E. Moore, "Progress in Digital Integrated Circuits," *Proceedings of the 1975 International Electron Devices Meeting*, pp. 11-13.

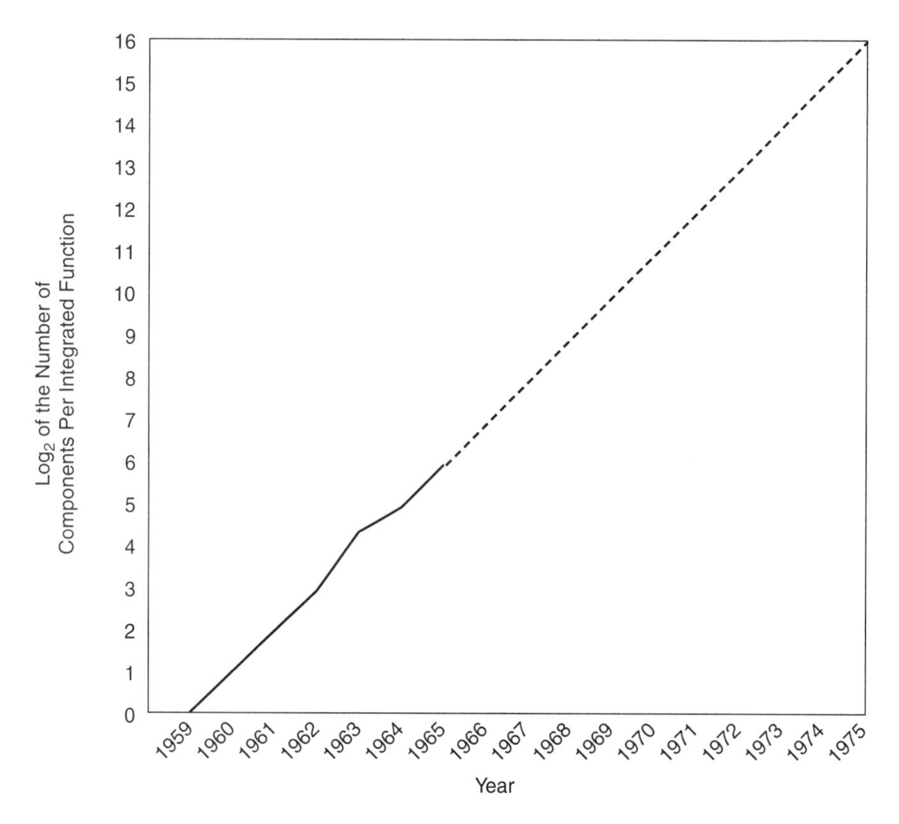

FIGURE 1 The original "Moore's Law" plot from *Electronics*, April 1965.

the number of transistors on a chip every 18 months.[10] While not deterministic, Moore's Law accurately reflects the pace for growth in the capacity of memory chips and logic chips from 1970 to 2002, as shown in Figure 2.[11]

[10]See Kenneth Flamm, "Moore's Law and the Economics of Semiconductor Price Trends," in National Research Council, *Productivity and Cyclicality in Semiconductors*, op. cit., for a comparison of Moore's predictions with the historical record. Flamm notes that Moore's own observations differ from what is popularly interpreted by the technology community and the press as Moore's Law. Though prescient, Moore did not anticipate the resilience of his earlier prediction. See Gordon E. Moore, "The Continuing Silicon Technology Evolution Inside the PC Platform," *Intel Developer Update*, Issue 2, October 15, 1997, where he notes that he "first observed the 'doubling of transistor density on a manufactured die every year' in 1965, just four years after the first planar integrated circuit was discovered. The press called it "Moore's Law," and the name stuck. To be honest, I did not expect this law to still be true some 30 years later, but now I am confident that it will be true for another 20 years."

[11]For a review of Moore's Law on its fortieth anniversary, see the *Economist*, "Moore's Law at 40," March 23, 2005.

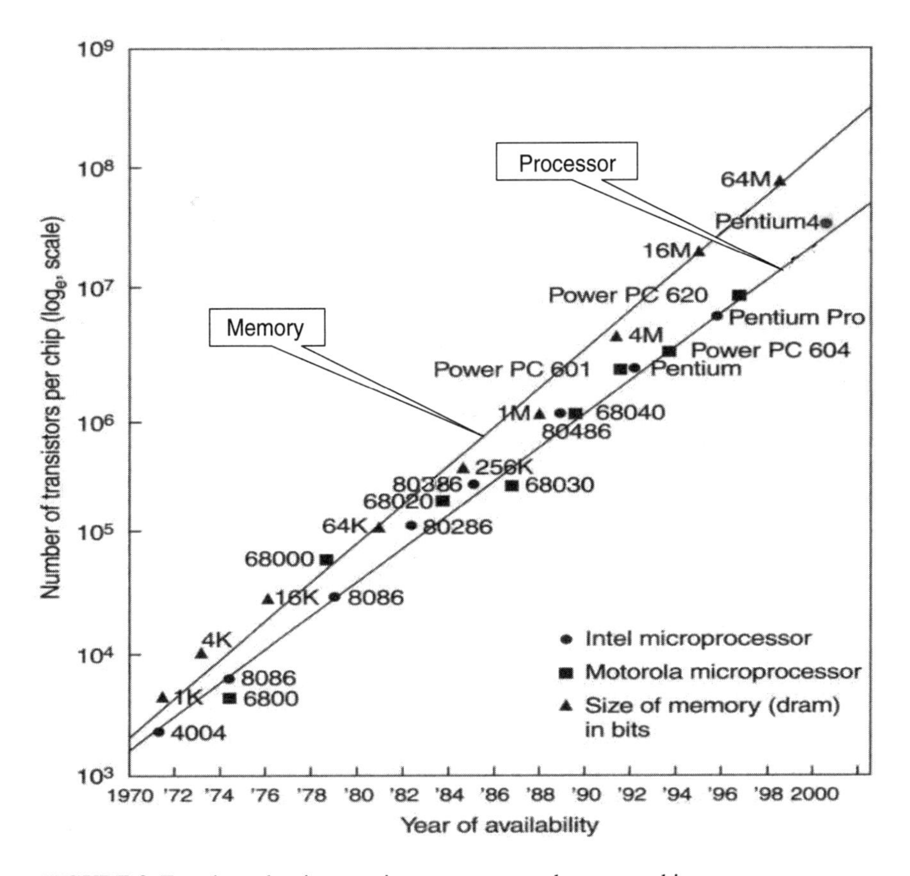

FIGURE 2 Transistor density on microprocessors and memory chips.

As Kenneth Flamm further noted, Moore's Law also captures an economic corollary that successive generations of semiconductors and related information technology products will not only be faster but also successively cheaper. Data from the Bureau of Economic Analysis (BEA), depicted in Figure 3 (and displayed by Dale Jorgenson at the conference on software), shows that quality-adjusted semiconductor prices have been declining by about 50 percent a year for logic chips and about 40 percent a year for memory chips between 1977 and 2000. This is unprecedented for a major industrial input.

The Moore's Law phenomenon also appears to extend from microprocessors and memory chips to high-technology hardware such as computers and communications equipment. BEA figures highlighted by Dale Jorgenson reveal also that computer prices have declined at about 15 percent per year since 1977. (See Figure 4.)

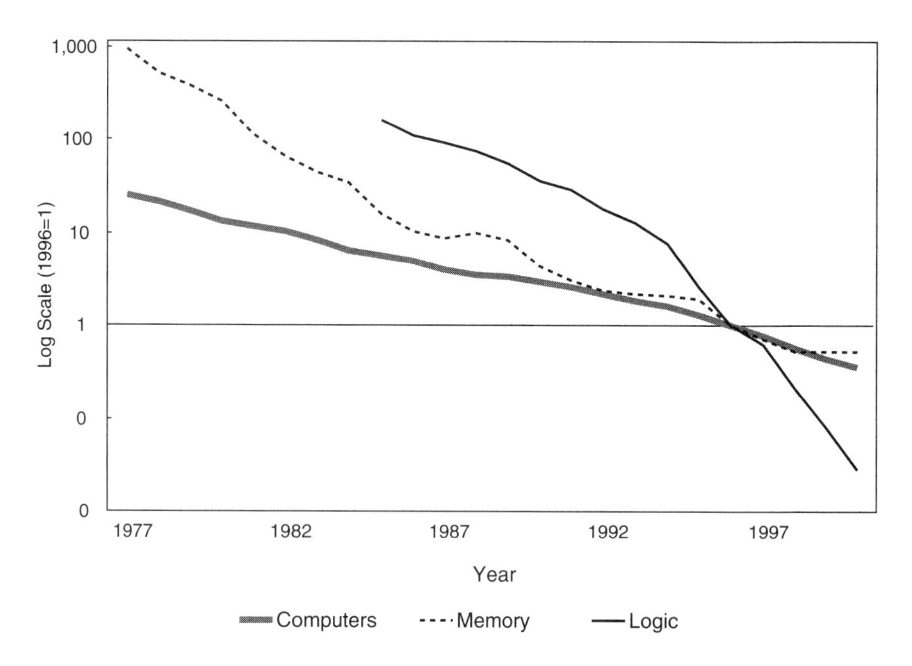

FIGURE 3 Relative prices of computers and semiconductors, 1977-2000.
NOTE: All price indexes are divided by the output price index.

While Moore's Law appears to predict ever "faster, better, cheaper" semiconductors and computers, it is not a deterministic law of nature, enduring instead by setting the expectations among participants in the semiconductor and computer industry of the pace of innovation and the introduction of new products to market. Before describing the basis of Moore's Law and what is required to sustain this remarkable phenomenon, we first summarize some of the discussion of the economic implications of Moore's Law and the challenges they pose to measuring the New Economy.

MEASURING THE NEW ECONOMY

Measuring the New Economy is a challenge given the fast-changing nature of information and communications technology and the complex and often-invisible roles it plays in economic processes. This means that current data collection methods have to be updated to stay relevant to new products, new categories, and new concepts.

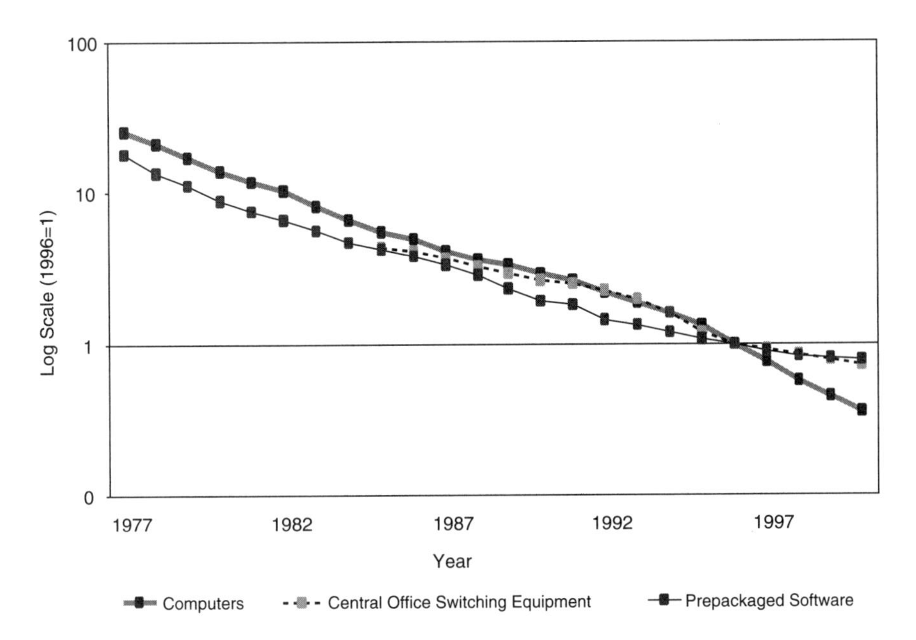

FIGURE 4 Relative prices of computers, communications, and software, 1977-2000.
NOTE: All price indexes are divided by the output price index.

The Challenge of Measurement

As several participants at the initial conference noted, conventional statistical methods are not adequately adapted to capture what is happening in the economy. Illustrating the challenges facing the federal statistical system, Timothy Bresnahan of Stanford University noted the discrepancy between measures of *output* in the information technology sector (which he noted are adequate) and measures of output where information technology is used as an *input* in other sectors (which are not).[12] Shane Greenstein of Northwestern University added that conventional measures of Gross Domestic Product (GDP) provide good data on established channels by which goods and services are distributed, but fail to capture such information about goods and services when there are concurrent changes in the distribution methods.[13]

Illustrating the implications of asymmetries in data availability, Lee Price (then of the Department of Commerce) observed that data on the value of pre-

[12]Tim Bresnahan, "Investments in Information Technology Applications," in National Research Council, *Measuring and Sustaining the New Economy*, op. cit., pp. 115-119.

[13]Shane Greenstein, "Communications," in National Research Council, *Measuring and Sustaining the New Economy*, op. cit., pp. 101-105.

Box A: Challenges in Measuring of the New Economy

Key challenges to measuring the New Economy—as noted by several of the participants at the initial conference on Measuring and Sustaining the New Economy—include:

- the need for better measurement of the output of the service sectors,
- the impact of technology on user sectors,
- the contribution of the semiconductor industry,
- the rapid changes in the communications industry,
- changes in distribution methods,
- organizational capital and other intangibles,
- assessing the value of business information systems, and
- difficulties when investments are reported as expenses in source data.

packaged software (which is more easily measured in terms of both nominal value and price) might not be as important to productivity as custom and own-account software whose value is more difficult to capture—resulting in their undervaluation. He stressed the need to refine statistical methods to better quantify the value of information technology.[14]

Several participants at the initial conference also emphasized the problems in *valuing* information technologies. Kenneth Flamm observed that it is difficult to calculate the percentage of improvement in computers that come from semiconductors.[15] Eric Brynjolfsson of the Massachusetts Institute of Technology (MIT) further noted some hazards in equating price with value for computers, particularly given that many consumers are not price-sensitive, valuing service, brand loyalty, and perceived quality instead.[16] Further to the issue of value, David Mowery of the University of California at Berkeley noted that it is statistically difficult to see the contributions of the semiconductor industry since it is hard to measure the output of "user" industries. He added that the economy outside the computer industry has become "a bit of a black planet" in terms of understanding quality improvements in its products.[17] This value issue was further elaborated at the conference on Deconstructing the Computer.

[14]This issue was pursued further in the 2003 National Academies conference on Software and the New Economy.

[15]Kenneth Flamm, "Microprocessors and Computers: The Phenomenon of Price Declines," in National Research Council, *Measuring and Sustaining the New Economy*, op. cit., pp. 82-89.

[16]Eric Brynjolfsson, "Economic Issues of E-Business," in National Research Council, *Measuring and Sustaining the New Economy*, op. cit., pp. 115-119.

[17]David Mowery, "Semiconductors, Economics of the New Economy," in National Research Council, *Measuring and Sustaining the New Economy*, op. cit., pp. 69-74.

Additional measurement challenges deal with how well information technologies are integrated and adopted across the economy. There were divergent views on where the United States was on the technology adoption curve at the turn of the century: Some argued that the United States was near the bottom of the S-shaped curve and about to take off; others suggested that the United States was in the middle and thus enjoying rapid productivity gains from the widespread adoption of information technologies; still others believed that marginal productivity gains from information technologies might be declining, signifying that the United States was already near the top of the technology adoption curve. This diversity of opinion, and the contrasting policy actions that it implies, pointed to a need to better measure the distinctive features of today's economy.

Another major constraint in sustaining the growth in productivity is the rate of technology absorption. Sid Abrams of AT Kearney noted that business organizations often face challenges in reengineering themselves to take better advantage of the technologies available. While the cutting edge of technologies may advance, their potential to advance business productivity may depend on the extent to which executives and others are aware of the possibilities and/or uncertain of the effects of adopting new technologies in their organization.[18] Indeed, as Ralph Gomery of the Alfred P. Sloan Foundation noted in the roundtable discussion that concluded the initial conference, the *ability to absorb* rapid advances in technology and the cost of re-doing the business organization to take advantage of these advances are, in many cases, more significant for sustaining productivity-led growth than the *rate of technological advance*. In essence, the question is not merely one of better or cheaper technology, but rather one of how enterprises can integrate productivity-enhancing technologies into the way business is conducted.

Sustaining the benefits of new technologies requires that we better understand the nature of these technologies and the circumstances that promote their development and deployment. STEP's series of conferences on the New Economy has thus sought to bring together leading economists and also to draw on the knowledge and experience of industry leaders and other experts to describe current trends and their origins, with the challenge to economists to identify data and tools required for measuring and modeling key facets of the New Economy.

Modeling the Productivity and Cyclicality of the Semiconductor Industry

Reflecting the centrality of semiconductors to the information technologies, STEP's conference of September 24, 2001, examined the rapid evolution of semiconductor technologies and a possible modeling strategy that could be used to predict the effects of alternative policy choices for the semiconductor industry.

[18]Sid Abrams, "Old Business to E-Business: The Change Dynamic," in National Research Council, *Measuring and Sustaining the New Economy*, op. cit., pp. 119-124.

Box B: Semiconductor Product Cyclicality and the New Economy

Intensifying competition in markets for semiconductor products
↓
Shift in product cycle for semiconductors from 3 to 2 years
↓
Sharp acceleration in price decline in semiconductors
↓
Substantial acceleration in information technology (IT) price declines, signaling faster productivity growth in IT-producing industries
↓
Powerful incentives for firms to substitute IT equipment for other forms of capital
↓
Boost in growth by nearly a full percentage point, with IT contributing more than half of this increase

SOURCE: Adapted from Dale Jorgenson, "Information Technology and the U.S. Economy," *American Economic Review*, 91(1):1-32, 2001.

Semiconductors and the New Economy

Participants at this conference noted that semiconductors are the basis of today's computing, information, and communications technologies. Rapid increases in the power of semiconductors, foreseen by Moore's Law, and corresponding rapid declines in the price of semiconductor-based information technologies have lead to their swift diffusion across the economy and propelled their adoption across an array of applications.[19]

Drawing on his 2001 presidential address to the American Economics Association, Dale Jorgenson reminded the conference participants that the resurgence in the U.S. economic growth trajectory since 1995 is associated with a "relentless" fall in semiconductor prices and coincident with a shift in product cycle for semiconductors from 3 to 2 years.[20] Jorgenson drew attention to a series of documented events—summarized in Box B—between an intensifying pace of competition in the market for semiconductor products and the boost in the aggregate growth rate of the U.S. economy.

[19]See Jeffrey T. Macher, David C. Mowery, and David A. Hodges, "Semiconductors," in *U.S. Industry in 2000: Studies in Competitive Performance*, David C. Mowery, ed., Washington, D.C.: National Academy Press, 1999, p. 245.

[20]See Dale W. Jorgenson, "Information Technology and the U.S. Economy," *American Economic Review*, 91(1):1-32, 2001.

Given that a disproportionate share of growth appears to be generated by increased efficiencies related to the production and use of information technology (IT), the economic consequences of a two-year product cycle—as opposed to a three-year product cycle—are significant. Jorgenson noted that the contribution of IT to growth from 1995 to 1999 was about 1.3 percent; by comparison, the annual growth of the U.S. economy over the same period was about 4 percent. A third of that is attributed to IT, meaning that 7 percent of the economy accounted for about a third of its economic growth. This is evidence, Jorgenson concluded, that the behavior of prices of IT, and the behavior of prices of semiconductors in particular, are of "momentous" importance to the economy.[21]

Explaining Productivity and Cyclicality in the Semiconductor Industry

Given its importance, how can economists better predict semiconductor price behavior? Participants at the conference highlighted the high sunk costs, steep learning curves, and rapid product cycles found in the semiconductor industry as factors affecting the industry's cyclicality. To predict price behavior, a successful industry model would have to take the effects of these features into account.

- ***High Sunk Costs***: Sunk costs are costs already incurred that cannot be recovered regardless of future events. In his conference presentation, Minjae Song of Harvard University noted that semiconductor firms face significant sunk costs in building and upgrading of new fabrication plants (often called "fabs") where a midsized fab today costs at least $1.5 billion to $2 billion to build. In addition, very large research and development (R&D) investments are required to enter this industry—typically as much as 10 to 15 percent of annual sales—with the R&D often specific to a particular market segment.[22]

- ***Steep Learning Curves***: Learning curves in semiconductor production are steep—approximately 70 percent. This means that a doubling of output drops unit costs by about 30 percent. In the semiconductor industry, however, these economies are not generated so much by greater labor productivity as by incremental changes to the automated technology. As Kenneth Flamm noted, improvements over the lifetime of a product's production come from more efficient *die shrinks*, which increase the chip density of a silicon wafer, and from *yield learning*, where the number of good chips on a wafer

[21]See the presentation by Dale W. Jorgenson, "Productivity and Growth: Alternative Scenarios," in National Research Council, *Productivity and Cyclicality in Semiconductors: Trends, Implications, and Questions*, op. cit., pp. 55-59.

[22]Minjae Song, "Semiconductor Industry," in National Research Council, *Productivity and Cyclicality in Semiconductors: Trends, Implications, and Questions*, op. cit., pp. 30-35.

increases over time as a percentage of the total number of chips that are manufactured.[23]

- ***Rapid Product Cycles:*** The semiconductor industry is distinctive in its continuous and rapid introduction of new generations of products (i.e., chips) and the dramatic difference in performance from one generation of product to the next. It is also characterized by very large R&D investments—typically as much as 10 to 15 percent of annual sales—with this R&D often specific to the segment of the market that the firm is entering. Over the past 10 years, the industry has produced five to six generations of semiconductors. When a firm puts a frontier product on the market, existing products become non-frontier. For example, when both the Pentium 2 and the Pentium 3 processors were on the market, the Pentium 3 was at the market frontier. Pentium 3 subsequently became the non-frontier product with the introduction of the Pentium 4 processor. According to Minjae Song, this rapid product cycling has meant that stocks of the current frontier product can quickly lose value with the introduction of the next-generation product.[24]

These features, taken together, affect the semiconductor industry's cyclicality. Conference participants described a variety of pathways in this regard:

- ***Drawing Down Inventories:*** Fast technological change in the semiconductor industry means that a semiconductor firm cannot reserve inventories as a way of smoothing out demand fluctuations if it hopes to remain competitive. Instead, given the short lifetimes of semiconductor products, firms expect that their inventories will lose value, even become obsolete, if held for too long. Considering the need to recoup high sunk costs, semiconductor firms face strong incentives to sell existing stocks of products as quickly as they can. This need to draw down inventories rapidly is thought to contribute to more pronounced industry cycles.[25]

- ***Excess Capacity:*** Attempts to capture the economies of the learning curve can also exacerbate the industry cycle. While, as noted above, the learning economies related to more efficient die shrinks and yield learning help cut costs, the hidden added capacity that results can also contribute to a chip

---

[23]Kenneth Flamm, "Economic Growth and Semiconductor Productivity," in National Research Council, *Productivity and Cyclicality in Semiconductors: Trends, Implications, and Questions,* op. cit., pp. 43-45.

[24]Minjae Song, "Semiconductor Industry," in National Research Council, *Productivity and Cyclicality in Semiconductors: Trends, Implications, and Questions,* op. cit.

[25]Kenneth Flamm, "Economic Growth and Semiconductor Productivity," in National Research Council, *Productivity and Cyclicality in Semiconductors: Trends, Implications, and Questions,* op. cit., pp. 43-45.

glut. Faced with excess supply, firms may have to close older fabrication facilities and/or lower prices.[26] These measures can add to the cyclicality of the industry.

- ***Time to Build:*** Finally, semiconductor fabrication plants take time to build—typically up to 2 years—and lag times between spikes in demand and sale can also play a significant role in the industry's cyclicality. Unanticipated surges in chip demand may be prompted by shocks such as those related to the mid-1990s boom in the PC market, the subsequent popularity of the Internet, and the rapid expansion (and later collapse) of the wireless communications market. Given that time is needed to build new manufacturing capacity, however, it is possible that demand fades just as the new capacity to meet this anticipated demand comes on stream. These lags between demand and supply, thus, can exacerbate cyclicality in the market for semiconductors.

In all, as David Morgenthaler of Morgenthaler Ventures observed at the conference, technological developments that decrease the cost per function and subsequently expand the depth and diversity of the market do not seem to translate into smoother industry cycles.

A Possible Model of the Semiconductor Industry

Models of the semiconductor industry that reflect its characteristic cyclicality can be a useful tool to predict semiconductor price behavior. In his conference presentation, Ariel Pakes of Harvard University described a modeling strategy that he has developed that he said can capture key features of complex and dynamic industries.[27] This model is based on "primitives" that determine each firm's profits conditional on the qualities of the products marketed, the costs of production, and the prices charged by all firms. This model could then be extended to include additional features of the specific industry being studied. Participants at the conference then examined the Pakes model to see if it could capture the salient features of the semiconductor industry.

A simple, static version of the Pakes model consists of a demand system, cost functions for each producer in the model, and an equilibrium assumption to solve reasonable pricing and quantity-setting decisions. Profits for each firm could then be calculated based on the price, the quality of each product sold, and the firm's cost function. The hope is that this type of model could be further extended to

---

[26]Ibid. Relatedly, see Jack Robertson, "Die Shrinks Now Causing Logic Chip Glut," *Semiconductor Business News*, October 15, 1998.

[27]See Ariel Pakes, "A Modeling Strategy for Industry," in National Research Council, *Productivity and Cyclicality in Semiconductors: Trends, Implications, and Questions*, op. cit., pp. 21-26.

consider some dynamic investment decisions that result from those profit estimates and their likely impact on the industry and on consumers.[28]

Deconstructing the Computer: Measuring Computer Hardware Performance

The next National Research Council (NRC) conference on the New Economy sought to deconstruct the computer into its components as a way of understanding its sources of growth and to discover how best to measure this growth. To this end, conference participants considered how the Moore's Law phenomenon of rapidly expanding capabilities applies to the various computer component industries.[29]

Although Gordon Moore's initial prediction pertained to changes in the semiconductor capacity, Moore's Law today more popularly captures the phenomenon of "faster" as well as "cheaper" development across a variety of computer components.[30] The conference brought together industrialists from leading computer hardware firms to explain how Moore's Law applied to their products and described the types of internal measures that industry had developed to track this change.

- ***Microprocessors:*** William Seigle of AMD, a microprocessor manufacturer, compared the Am386, introduced by his company in 1991, with the Opteron, introduced in April 2003. Performance, he noted, had jumped 50 times from 33 MHz to 2 GHz, offering significant improvements in the efficiency of instruction processing, memory hierarchy, and branch prediction.[31]

- ***Hardware Storage:*** Remarking on the performance improvements in computer storage, Robert Whitmore of Seagate Inc. noted that performance, measured as input/output transactions per second, had accelerated significantly between the late 1980s and late 1990s. Meanwhile, he noted that the price

[28]C. Lanier Benkard of Stanford University illustrated how Dr. Pakes' framework has been adapted to model the U.S. aircraft industry. See C. Lanier Benkard, "The Case of the Aircraft Industry," in National Research Council, *Productivity and Cyclicality in Semiconductors: Trends, Implications, and Questions*, op. cit., pp. 26-30.

For additional detail, see C. Lanier Benkard, *A Dynamic Analysis of the Market for Wide Bodied Commercial Aircraft*, Graduate School of Business, Stanford University, June 2001.

[29]The nature of Moore's Law is described later in this chapter in the section on "Sustaining the New Economy."

[30]Kenneth Flamm, "Economic Growth and Semiconductor Productivity," in National Research Council, *Productivity and Cyclicality in Semiconductors: Trends, Implications, and Questions*, op. cit., pp. 43-45.

[31]William Seigle, "Processor Evolution," in National Research Council, *Deconstructing the Computer*, Dale W. Jorgenson and Charles W. Wessner, eds., Washington, D.C.: The National Academies Press, 2005.

of rotating magnetic memory on a dollar-per-gigabyte basis had eroded at an annual compound rate of *minus* 45 percent between 1995 and 2002. In addition, mean time between failures, a measure of reliability, had grown at a phenomenal compound annual rate of 25 percent from 1977 to 2001.[32]

- ***Software Storage Systems:*** Mark Bregman of Veritas Software (a company that develops software to help store, access, and manage data) noted the apparent observance of Gilder's Law, which states that the total bandwidth of communication systems triples every 12 months.[33] Further, he noted that storage devices achieve 100 percent growth in density annually, a reality that translates into better cost at a dramatic rate.[34]

- ***Graphics:*** Chris Malachowsky of NVIDIA documented product performance improvements in graphics from the second half of 1997 to the first half of 2003 at an annualized rate of 215 to 229 percent. Rapid technological advances in graphics technology, he noted, rendered moviemaking chores, previously requiring farms of thousands of machines, to be possible using consumer PCs, dramatically lowering prices.[35]

Developing Hedonic Price Indexes

Several participants at the conference on computers emphasized the need to develop appropriate categories and performance measures to capture the growth of these dynamic and complex industries. The Brookings Institution's Jack Triplett underscored this point in his conference presentation, emphasizing that economists need to learn more about the contributions of hardware component technologies to the increase in computer performance.[36]

Dr. Triplett noted that while the cost of computing today is projected to be about one-thousandth of one percent of what it cost 50 years ago, this estimate still does not account for all aspects of computer performance. An exciting

---

[32]Robert Whitmore, "Storage," in National Research Council, *Deconstructing the Computer*, op. cit.

[33]Aron, Dunmore, and Pampush estimate 85 percent per annum growth in bandwidth since 1983. See Debra J. Aron, Ken Dunmore, and Frank Pampush, "Worldwide Wait? How the Telecom Act's Unbundling Requirements Slow the Development of the Network Infrastructure," *Industrial & Corporate Change,* 7(4):615-621, 1998.

[34]Mark Bregman, "The Promise of Storage Systems," in National Research Council, *Deconstructing the Computer*, op. cit.

[35]Chris Malachowsky, "Graphics," in National Research Council, *Deconstructing the Computer*, op. cit.

[36]Jack Triplett, "Performance Measures for Computers," in National Research Council, *Deconstructing the Computer,* op. cit. This paper provides a general overview of the scope and limitations of the Hedonic methodology for computers.

research agenda, he noted, is to account for the determinants of the great decline in computer price/performance over the last 50 years.

This research agenda is very challenging because it has to account for the qualitative changes in computers. How do we measure, for example, the performance of the computer and its components through time? Dr. Triplett acknowledged that the question is complicated by the dynamism and complexity of the technological change characterizing the evolution of the modern computer. To be sure, the cost, capabilities, and size of a 1952 UNIVAC are significantly different from those of a modern laptop. Since direct comparisons of price are not feasible—the proverbial apples and oranges problem—a key challenge for economists is to adjust their price data for quality differences. Indeed, identifying such "true price change" has long been a goal of price statisticians and national accountants.

One way of adjusting prices for quality differences is to use hedonic price indexes. Developed 40 years ago by Zvi Griliches and enhanced since, this econometric method takes into account an array of characteristics possessed by a product and their functional relation to price.[37] Many economists regard hedonic price indexes to be a theoretically promising way of adjusting for quality when measuring the price of computing power through time, while recognizing the need for further development.[38]

Methodological Challenges and Opportunities for Hedonic Pricing

In practice, however, the continued dynamism and complexity of the relevant industries will make the task of developing robust measures of computer performance highly challenging. Rapid supply-driven evolution of products and concepts, as well as changing consumer behavior, keeps the industry in flux, rendering the economist's task more difficult. Swift technological change can change and, in some cases, even make obsolete the relative importance of particular quality characteristics used in hedonic estimates.[39] A further prob-

[37]Zvi Griliches, "Hedonic Price Indexes for Automobiles: An Econometric Analysis of Quality Change," in G. Stigler (chairman), *The Price Statistics of the Federal Government,* New York, NY: Columbia University Press, 1961.

[38]A National Academies panel has noted that "Hedonic techniques currently offer the most promising approach for explicitly adjusting observed prices to account for changing product quality. But our analysis suggests that there are substantial unresolved econometric, data, and other measurement issues that need further attention." National Research Council, *At What Price? Conceptualizing and Measuring Cost-of-Living and Price Indexes,* Washington D.C.: The National Academies Press, 2002, Chapter 4.

[39]For example, the hedonic methodology used by BEA to estimate quality-adjusted microprocessor prices in the period up to 1996 could not have easily been extended into a later time period. Nearly all of the quality characteristics—other than speed—were present in nearly all of the microprocessor chips at the end of the sample period. Further, although the earliest Pentium chips were available near the end of the sample, the methodology used was unable to capture some of the improvements in computing power brought on by replacing 486-generation chips with Pentium Is.

lem arises because, if things change enough, no price methodology will give accurate estimates.[40]

Illustrating this technological dynamism, Dalen Keyes of DuPont Displays noted that the U.S. display industry sees its future in moving away from LCD (liquid crystal display) technologies and towards Organic LED (light-emitting diode) technologies. OLED display technology, based on a roll-to-roll manufacturing concept, integrates components from the flex-circuitry industry with inkjet printing from the graphic arts industry to get rolls of material that could be "sliced and diced" into displays. Flexible and versatile, OLEDs, he predicted, will possess qualities and applications quite different from today's displays.[41]

Tracing the evolution of technology in the printer industry, Howard Taub of Hewlettt-Packard noted that "we are pretty much at a point where the quality of the image that you can print is about as good as you're going to get." As a result, he noted, the quest for "better" had gone on to pursue other dimensions including connectivity and ease of use. He also noted that the computer printer industry is looking to create new markets beyond those for office printing and duplication. New printer technologies, he noted, could enable the production of limited-run custom magazines and advertisements, changing the way consumers think about desktop printers.[42]

Indeed, for displays and printers, as with other computer components, the use of hedonic indexes to control for quality of a product is likely to be a challenge as continuing rapid innovation changes not just the features of the product but even the concept of the product itself.

Another challenge to developing robust hedonic price indexes arises when—as David McQueeney of IBM put it—"faster, better, cheaper," collides with the "good enough phenomenon."[43] He noted, for example, that many current models of displays and home computers have crossed the "good enough" threshold for most of today's home computing needs—the point also raised by Dr. Taub, above. Displays used for everyday desktop home-PC applications have become so good, observed Dr. McQueeney, that "further technological improvements aimed at more pixels per inch could not be detectable to the end user." Similarly, he noted that disk capacity has become so large that most ordinary users never fill the hard

[40]An example of this is Robert Gordon's "Hulten-Breugel paradox," which notes that extending price estimates back to late medieval times results in German peasants living on virtually nothing in real terms, yet Breugel's paintings show them as well-housed, well-clothed, and well-fed. See Robert Gordon, "Apparel Prices and the Hulten-Breugel Paradox," paper presented at the CRIW Conference on Price Index Concepts and Measurement, October 15, 2004.

[41]Dalen Keys, "Flat Panel Displays," in National Research Council, *Deconstructing the Computer,* op. cit.

[42]Howard Taub, "Laser and Ink Jet Printers," in National Research Council, *Deconstructing the Computer*, op. cit.

[43]David McQueeney, "Overview of the IBM Global Product Plan," in National Research Council, *Deconstructing the Computer*, op. cit.

drive in the 2 or 3 years they normally keep a computer. So, although research and development do not stop at a certain point, their benefits may begin to show up in price reduction and cost-performance reduction rather than in performance measures. "The raw capabilities of technology have in some cases gotten to the point where either the economics of how you sell them and how you ascribe value to them is changing," he explained, "or you are forced to look elsewhere in the system performance stack to get real improvements."[44]

Dr. McQueeney also noted that the value of "faster and better" might remain unrealized pending additional developments in technology and finance. Looking ahead to the conference on the Telecommunications Challenge, he observed that there is at present enough fiber capacity to "connect every person in North America to every person in Eastern and Western Europe and to allow all to have a phone conversation at the same time." He also noted that a tremendous capacity in optical fiber has been installed between various cities and within metropolitan areas of the United States. Yet, "the intelligence needed to light up those fiber-optic networks and make them actually do something useful—the servers, the routers, the switches—is in fact quite expensive," he stated, "and we're still struggling with a good investment model that will let us build out that control infrastructure to use the fiber capacity we have."[45] This need to realize necessary complementarities was also echoed by Dr. Siegle, who noted that "while microprocessors are important, you can't make meaningful systems and applications if there are advances in just the microprocessor."[46]

These conceptual challenges to measuring performance aside, industry experts at the conference described a variety of formal and informal measures currently used by computer component industries to gauge performance. Dr. Whitmore noted that for the hardware storage industry, capacity in bytes, price, performance, and reliability remain the main factors for measurement, although additional metrics are appearing on the horizon. In the printer industry, "faster and better" is measured in terms of printer speed, resolution, reliability, and usability, according to Dr. Taub. Mr. Malachowsky noted that there is a marketing view of performance in addition to internal and external views in the graphics industry. He noted that his company, NVIDIA, measures itself internally on "very engineering-specific, design-specific things" such as bandwidth utilization factors and externally according to particular application benchmarks. Echoing the common theme, Dr. Keyes noted that the display industry relies on an extensive list of technical specifications, including diagonal size of the display, pixel count, and power consumption. Performance measures include luminants

[44]Ibid.

[45]Ibid. This point was further developed in the NRC conference on the *Telecommunications Challenge: Changing Technologies and Evolving Policies,* which is a part of the New Economy series.

[46]William Seigle, "Processor Evolution," in National Research Council, *Deconstructing the Computer*, op. cit.

and switching speed, which indicates whether a product could do video-grade displays. Another metric, he added, is the size of the substrate used in the manufacture displays.

Citing these and other performance measures made note of by the participants, Dr. Jorgenson concluded that measuring progress in the computer and computer component industries is not only possible, but that such measurement is increasingly more sophisticated and, in fact, "quite successful." He recalled that a set of measures for computers and peripherals begun in the late 1960s—grounded in economics research at IBM—achieved incorporation into the U.S. national accounts for the first time in the mid-1980s. These have continued to be in use (while also being enhanced and developed) to the present day. He expressed optimism that similar progress on data measurement and analysis can be made based on what he had heard at this conference—and that this could help improve the economic understanding needed to develop the policies necessary to sustain the New Economy.[47]

Measuring Software Performance

Within the U.S. national accounts, software is broken down into three categories: prepackaged, custom, and own-account software. *Prepackaged* (or shrink-wrapped) software is packaged, mass-produced software. It is available off-the-shelf, though increasingly replaced by on-line sales and downloads over the Internet. In 2003, BEA placed business purchases of prepackaged software at around $50 billion. *Custom* software refers to large software systems that perform business functions such as database management, human resource management, and cost accounting.[48] In 2003, BEA estimated business purchases of custom software at almost $60 billion. Finally, *own-account* software refers to software systems built for a unique purpose, generally a large project such as an airlines reservation system. In 2003, BEA estimated business purchases of own-account software at about $75 billion.[49]

Dr. Jorgenson, in introducing the New Economy conference on Software, noted that while there is sufficient price information on prepackaged software, this category is only thought to make up about 25 to 30 percent of the software

[47]Dale W. Jorgenson, "Concluding Remarks," in National Research Council, *Deconstructing the Computer*, op. cit.

[48]The line between prepackaged and custom software is not always distinct. National accountants have to determine, for example, whether Oracle 10i, which is sold in a product-like fashion with a license, is to be categorized as custom or prepackaged software.

[49]David Wasshausen, "A BEA Perspective: Private Fixed Software Investment," in National Research Council, *Software, Growth, and the Future of the U.S. Economy,* Charles W. Wessner, ed., Washington, D.C.: The National Academies Press, 2006.

market.[50] Consequently, he noted, "there is a large gap in our understanding of the New Economy."[51]

Measurement Challenges: The Complexity of Software

Before we can develop appropriate measures of software performance, we first need to understand the nature of software itself. As William Raduchel of the Ruckus Network explained at the conference on software, software comprises millions of lines of code, operated within a *stack*.[52] The stack begins with the *kernel*, which is a small piece of code that talks to and manages the hardware. The kernel is usually included in the *operating system*, which provides the basic services and to which all programs are written. Above this operating system is *middleware*, which "hides" both the operating system and the window manager. For the case of desktop computers, for example, the operating system runs other small programs called *services* as well as specific *applications* such as Microsoft Word and PowerPoint.

Thus, when a desktop computer functions, the entire stack is in operation. This means that the value of any part of a software stack depends on how it operates within the context of the rest of the stack.[53] The result, as Monica Lam of Stanford University suggested, is that software may be the most intricate thing that humans have learned to build. Software grows more complex as more and more lines of code accrue to the stack, making software engineering much more difficult than other fields of engineering.[54]

The way software is written also adds to its complexity and cost. As Anthony Scott of General Motors pointed out, the process by which corporations build software is "somewhat analogous to the Winchester Mystery House," where accretions to the stack over time create a complex maze that is difficult to fix or change.[55]

[50]A weakness of the official price estimates for custom and own-account software is that they are not well adjusted for quality change. The Bureau of Economic Analysis has contracted with a private firm to produce improved price indexes for custom software using hedonic methods and a number of functional characteristics as explanatory variables. If this work is successful, it will likely lead to more rapid price declines for custom—and by extension, own-account—software.

[51]Dale Jorgenson, "Introduction," in National Research Council, *Software, Growth, and the Future of the U.S. Economy*, op. cit.

[52]William Raduchel, "The Economics of Software," in National Research Council, *Software, Growth, and the Future of the U.S. Economy*, op. cit.

[53]Other IT areas have their own idiomatic "stack" architectures. For example, there are more CPUs in industrial control systems than on desktops, and these embedded systems do not have "window managers." A similar point can be made for mainframe systems, distributed systems, and other non-desktop computing configurations.

[54]Monica Lam, "How Do We Make It?" in National Research Council, *Deconstructing the Computer*, op. cit.

[55]The Winchester Mystery House, in San Jose, California, was built by the gun manufacturer heiress who believed that she would die if she stopped construction on her house. Ad hoc construction,

This complexity means that a failure manifest in one piece of software, when added to the stack, may not indicate that something is wrong with that piece of software *per se*, but quite possibly can cause the failure of some other piece of the stack that is being tested for the first time in conjunction with the new addition.[56] In short, the complexity of software makes measuring software performance very challenging.

Tracking Software in National Accounts

The unique nature of software also poses challenges for national accountants who are interested in data that track software costs and aggregate investment in software and its impact on the economy. This is important because over the past 5 years, investment in software has been about 1.8 times as large as private fixed investment in computers' peripheral equipment and was about one-fifth of all private fixed investment in equipment and software.[57] Getting a good measure of this asset, however, is difficult because of the unique characteristics of software development and marketing, as well as the conventions by which it is reported.

According to Shelly Luisi of the Securities and Exchange Commission (SEC), some data about software come from information that companies report to the SEC.[58] These companies follow the accounting standards developed by the Financial Accounting Standards Board (FASB).[59] Luisi noted that the FASB developed these accounting standards with the investor, and not a national accountant, in mind. As a result of these accounting standards, she noted, software is included as property, plant, and equipment in most financial statements rather than as an intangible asset.[60]

---

starting in 1886 and continuing over nearly four decades with no master architectural plan, created an unwieldy mansion with a warren of corridors and staircases that often lead nowhere.

[56]Anthony Scott, "The Role of Software," in National Research Council, *Deconstructing the Computer*, op. cit.

[57]Bureau of Economic Analysis, National Income and Product Income, Table 5.3.5 on Private Fixed Investment by Type.

[58]Currently, the data estimates for annual estimates of prepackaged and custom software (in current dollars) come from Census annual services surveys that are benchmarked to the quinquennial Input-Output tables. These tables, in turn, incorporate information from quinquennial economic censuses. Annual own-account software is based primarily on Bureau of Labor Statistics (BLS) estimates of numbers of programmers and computer systems analysts, plus salaries per year for same, plus overhead costs. Only the first two (of three current quarterly) estimates of a quarter's software investment make use of data reported to the SEC. Currently, therefore, the overall picture of software investment is not entirely dependent on the SEC data.

[59]The Financial Accounting Standards Board (FASB) is a private organization that establishes standards of financial accounting and reporting governing the preparation of financial reports. They are officially recognized as authoritative by the Securities and Exchange Commission.

[60]Outlining the evolution of the FASB's standards on software, Ms. Luisi recounted that the FASB's 1974 Statement of Financial Accounting Standards (FAS-2) provided the first standard for capitalizing software on corporate balance sheets. FAS-2 has since been developed though further interpretations

Given these accounting standards, how do software companies actually recognize and report their revenue? Taking the perspective of a software company, Greg Beams of Ernst & Young noted that while sales of prepackaged software are generally reported at the time of sale, more complex software systems require recurring maintenance to fix bugs and to install upgrades, causing revenue reporting to become more complicated. In light of these multiple deliverables, software companies come up against rules requiring that they allocate value to each of those deliverables and then recognize revenue in accordance with the requirements for those deliverables. How this is put into practice results in a wide difference in when and how much revenue is recognized by the software company, he noted—making it, in turn, difficult to understand the revenue numbers that a particular software firm is reporting.[61]

Mr. Beams noted that information published in software vendors' financial statements is useful mainly to the shareholder. He acknowledged that detail is often lacking in these reports, and that distinguishing one software company's reporting from another and aggregating such information so that it tells a meaningful story can be extremely challenging.

Gauging Private Fixed Software Investment

Although the computer entered into commercial use some four decades earlier, the Bureau of Economic Analysis has recognized software as a capital investment (rather than as an intermediate expense) only since 1999. Describing BEA methodology, David Wasshausen of BEA noted that his organization uses a "commodity flow" technique to measure prepackaged and custom software. Beginning with total receipts, BEA adds imports and subtracts exports, which leaves the total available domestic supply. From that figure, BEA subtracts household and government purchases to come up with an estimate for aggregate business investment in software.[62] By contrast, BEA calculates own-account software

and clarifications. FASB Interpretation No. 6, for instance, recognized the development of software as R&D and drew a line between software for sale and software for operations. In 1985, FAS-86 introduced the concept of technological feasibility, seeking to identify that point where the software project under development qualifies as an asset, providing guidance on determining when the cost of software development can be capitalized. In 1998, FASB promulgated "Statement of Position 98-1" that set a different threshold for capitalization for the cost of software for internal use—one that allows it to begin in the design phase, once the preliminary project state is completed and a company commits to the project. Shelly Luisi, "Accounting Rules: What do they Capture and What are the Problems?" in National Research Council, *Software, Growth, and the Future of the U.S. Economy,* op. cit.

[61]Greg Beams, "Accounting Rules: What do they Capture and What are the Problems?" in National Research Council, *Software, Growth, and the Future of the U.S. Economy,* op. cit.

[62]David Wasshausen, "A BEA Perspective: Private Fixed Software Investment," in National Research Council, *Software, Growth, and the Future of the U.S. Economy,* op. cit. BEA compares demand-based estimates for software available from the U.S. Census Bureau's Capital Expenditure Survey with the supply-side approach of the commodity flow technique. The Census Bureau is

as the sum of production costs, including compensation for programmers and systems analysts and such intermediate inputs as overhead, electricity, rent, and office space.[63]

According to Dr. Wasshausen, BEA is striving to improve the quality of its estimates. While BEA currently bases its estimates for prepackaged and custom software on trended earning data from corporate reports to the SEC, it hoped to benefit soon from Census Bureau data that capture receipts from both prepackaged and custom software companies through quarterly surveys. Among recent BEA improvements, Dr. Wasshausen cited an *expansion of the definitions* of prepackaged and custom software imports and exports, and *better estimates* of how much of the total prepackaged and custom software purchased in the United States was for intermediate consumption. BEA, he said, was also looking forward to an improved Capital Expenditure Survey by the Census Bureau.[64]

Dirk Pilat of the Organisation for Economic Co-operation and Development (OECD) noted at the same conference that methods for estimating software investment have been inconsistent across the countries of the OECD.[65] One problem contributing to the variation in measures of software investment is that the computer services industry represents a heterogeneous range of activities, including not only software production, but also such things as consulting services. National accountants have had differing methodological approaches (for example, on criteria determining what should be capitalized) leading to differences between survey data on software investment and official measures of software investments as they show up in national accounts.

Attempting to mend this disarray, Dr. Pilat noted that the OECD Eurostat Task Force has published its recommendations on the use of the commodity flow model and on how to treat own-account software in different countries.[66] He noted that steps were under way in OECD countries to harmonize statistical

---

working to expand its survey to include own-account software and other information not previously captured, according to David Wasshausen.

[63]BEA's estimates for own-account are derived from employment and mean wage data from the BLS's Occupational Employment Wage Survey and a ratio of operating expenses to annual payroll from the Census Bureau's Business Expenditures Survey.

[64]David Wasshausen, "A BEA Perspective: Private Fixed Software Investment," in National Research Council, *Software, Growth, and the Future of the U.S. Economy,* op. cit.

[65]Dirk Pilat, "What is in the OECD Accounts and How Good is it?" in National Research Council, *Software, Growth, and the Future of the U.S. Economy,* op. cit. Countries that ask sellers of software, "How much did you sell?" find that there is a lot more software investment than do the countries that ask the buyers of software "How much did you buy?" The Bureau of Economic Analysis analyzed data based on both questions, and found that the "sell" question estimates—which underlie the published estimates—yielded estimates roughly an order of magnitude larger. The published estimates are adjusted for non-software production activities.

[66]Organisation for Economic Cooperation and Development, *Statistics Working Paper 2003/1: Report of the OECD Task Force on Software Measurement in the National Accounts*, Paris: Organisation for Economic Co-operation and Development, 2003.

Box C: The Economist's Challenge: Software as a Production Function

Software is "the medium through which information technology expresses itself," says William Raduchel. Most economic models miscast software as a machine, with this perception dating to the period, 40 years ago, when software was a minor portion of the total cost of a computer system. The economist's challenge, according to Dr. Raduchel, is that software in not a factor of production like capital and labor, but actually embodies the production function, for which no good measurement system exists.

practices and that the OECD would monitor the implementation of the Task Force recommendations. This effort would then make international comparisons possible, resulting in an improvement in our ability to ascertain what was moving where—the "missing link" in addressing the issue of offshore software production.

Despite the comprehensive improvements in the measurement of software undertaken since 1999, Dr. Wasshausen noted that accurate software measurement continued to pose severe challenges for national accountants simply because software is such a rapidly changing field. He noted, in this regard, the rise of demand computing, open-source code development and overseas outsourcing, which create new concepts, categories, and measurement challenges.[67] Characterizing attempts made so far to deal with the issue of measuring the New Economy as "piecemeal"—"we are trying to get the best price index for software, the best price index for hardware, the best price index for LAN equipment routers, switches, and hubs"—he suggested that a single comprehensive measure might better capture the value of hardware, software, and communications equipment in the national accounts. Indeed, information technology may best be thought of as a "package," combining hardware, software, and business-service applications.[68]

[67]For example, how is a distinction to be made between service provisioning (sending data to a service outsource) and the creation and use of a local organizational asset (sending data to a service application internally developed or acquired)? The user experience may be identical (e.g., web-based access) and the geographic positioning of the server (e.g., at a secure remote site, with geography unknown to the individual user) may also be identical. In other words, the technology and user experience both look almost the same, but the contractual terms of provisioning are very different.

[68]David Wasshausen, "A BEA Perspective: Private Fixed Software Investment," in National Research Council, *Software, Growth, and the Future of the U.S. Economy,* op. cit.

Tracking Software Price Changes

A further challenge in the economics of software lies in tracking price changes. Drawing on Microsoft Corporation data, Alan White of Analysis Group and Ernst Berndt of MIT presented their work on estimating price changes for prepackaged software.[69] Dr. White noted that an investigator faces several important challenges in constructing measures of price and price change. These include ascertaining which price to measure because software products may be sold as full versions or as upgrades, stand-alones, or suites. An investigator has also to determine what the unit of output is, how many licenses there are, and when price is actually being measured. Another key issue, he added, concerns how the quality of software has changed over time and how that should be incorporated into price measures.[70]

Surveying the types of quality changes that might come into consideration, Dr. Berndt gave the example of improved graphical interface and "plug-'n-play," as well as increased connectivity between difference components of a software suite.[71] Referring to their study, Dr. Berndt noted that he and Dr. White compared the *average* price level (computing the price per operating system as a simple average) with *quality-adjusted* prices levels using hedonic and matched-model econometric techniques. They found that while the average price, which does not correct for quality changes, showed a growth rate of about 1 percent a year, the quality-adjusted matched model showed a price decline of around 6 percent a year and the hedonic calculation showed a much larger price decline of around 16 percent.

These quality-adjusted price declines for software operating systems, shown in Figure 5, support the general thesis that improved and cheaper information technologies contributed to greater information technology adoption leading to productivity improvements characteristic of the New Economy.[72]

Measuring Telecom Prices

How do new information and communications technologies translate into prices and hence consumer welfare? Mark Doms of the Federal Reserve Bank of San Francisco provided the participants in the STEP conference on the Tele-

[69]Jaison R. Abel, Ernst R. Berndt, and Alan G. White, "Price Indexes for Microsoft's Personal Computer Software Products," NBER Working Paper 9966, 2003. The research was originally sponsored by Microsoft Corporation, though the authors are responsible for its analysis.

[70]Alan White, "Measuring Prepackaged Software," in National Research Council, *Software, Growth, and the Future of the U.S. Economy,* op. cit.

[71]Ernst Berndt, "Measuring Prepackaged Software," in National Research Council, *Software, Growth, and the Future of the U.S. Economy,* op. cit.

[72]Dale W. Jorgenson and Kevin J. Stiroh, "Raising the Speed Limit: U.S. Productivity Growth in the Information Age," Brookings Papers on Economic Activity, Washington, D.C.: The Brookings Institution, 2000.

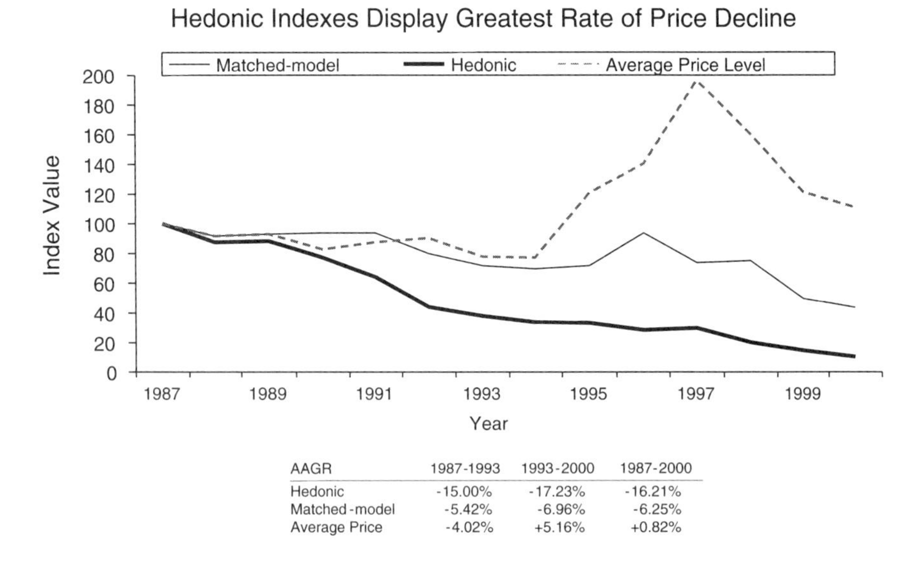

AAGR	1987-1993	1993-2000	1987-2000
Hedonic	-15.00%	-17.23%	-16.21%
Matched-model	-5.42%	-6.96%	-6.25%
Average Price	-4.02%	+5.16%	+0.82%

FIGURE 5 Quality-adjusted prices for operating systems have fallen, 1987-2000.
SOURCE: Jaison R. Abel, Ernst R. Berndt, Cory W. Monroe, and Alan White, "Hedonic Price Indexes for Operating Systems and Productivity Suite PC Software," draft working paper, 2004.

communications Challenge an overview of what the current official numbers say, and the challenges of coming up with good price indexes for communications equipment and services. He noted that while investment in communications in the United States had been substantial—around $100 billion per year, representing a little over 10 percent of total equipment investment in the U.S. economy—it had also been highly volatile. During the recession of the early 2000s, he noted, IT investment fell about 35 percent from peak to trough (see Figure 6[73]). Dr. Doms noted that this recession might well be remembered as the high-tech recession, adding that "certainly what happened to communications played a major role in what happened to the high-tech sector."

Measuring the dollars spent on communications in the United States every year is difficult because technology is rapidly changing. As we noted earlier, a computer costing a thousand dollars today is a lot more powerful and versatile than a similarly priced one of 10 years ago—and this improvement is no less true for communications equipment. Similarly, most long-distance communications 25 years ago was handled through landline phones, in stark contrast to the diver-

[73]Mark Doms, "The Boom and Bust in Information Technology Investment," *Federal Reserve Bank of San Fransisco Economic Review*, 2004, pp. 19-34.

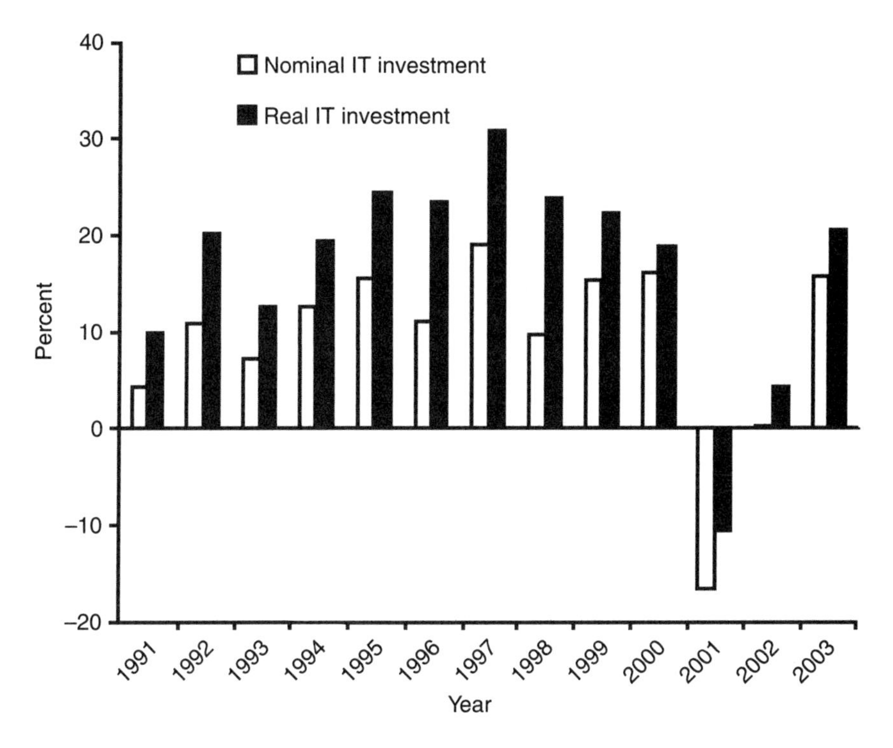

FIGURE 6 Annual percent change in IT investment.
SOURCE: Bureau of Economic Analysis.
NOTE: Percent changes based on year-end values.

sity of means of communications in use today. The technology is also in rapid flux. Dr. Doms noted that between 1996 and 2001 alone, there were tremendous advances in the amount of information that could travel down a strand of glass fiber, adding that the price of gear used to transmit information over fiber fell, on average, by 14.9 percent a year over this five-year period.

The fast speed of technological change renders the job of tracking prices (which enables us to see how much better off society is as a result of technological changes) a complex one. Whereas money spent on telecommunications was relatively easier to track 25 years ago when most purchases were of telephone switches, today's telecommunications equipment includes a wide array of technologies related to data, computer networking, and fiber optics.

Current methodologies for making inter-temporal comparisons in price and quality understate true price declines because they do not fully track these technological changes. While BEA has estimated that prices for communications gear fell an average of 3.2 percent per year between 1994 and 2000—in sharp contrast to the 19.3 percent fall in computer prices—Dr. Doms noted that more a complete estimate that he had developed shows that communications equipment prices actually fell on the order of 8 to 10 percent over that period.[74]

Towards Improved Measures of the New Economy

While this new estimate is a step in the right direction, Dr. Doms acknowledged that more refinement is necessary in measuring telecom prices. Echoing a refrain heard at each of the conferences in the series on Measuring and Sustaining the New Economy, he noted that the job of keeping track of rapid developments in information and communications technologies was growing increasingly difficult for statistical agencies, especially in light of their limited budgets and the rapid development of technology. "Unless the statistical agencies get increased funding, in the future, they are not going to be able to follow new, evolving trends very well," he concluded.

SUSTAINING THE NEW ECONOMY

The second theme of the NRC conferences on the New Economy concerned public polices needed to sustain the New Economy. A major focus of these conferences was on polices to sustain Moore's Law, the driver of faster and more widely affordable computers and other productivity-enhancing technologies. Participants at the conferences on software and telecommunications also examined the new challenges in globalization emerging from the possibility of sending voice and data at very low costs around the world.

To be sure, the challenge of measuring the New Economy and policies needed to sustain the benefits of the New Economy are two sides of one coin. Better data on what is moving where in offshoring are likely to permit more informed policy debate.

Challenges to Sustaining Moore's Law

As noted at the outset, Moore's Law is not a deterministic law but a self-fulfilling prophecy that needs to be sustained if the economy is to continue to benefit from the advantages of faster and cheaper information technologies.[75] Moore's Law works by setting expectations about the pace of competition in the semiconductor industry. Each firm, believing its rivals to develop and market

[74]Mark Doms, "Communications Equipment: What Has Happened to Prices?" Federal Reserve Bank of San Fransisco Working Paper 2003-15, 2003.

[75]Kenneth Flamm, "Moore's Law and the Economics of Semiconductor Price Trends," in National Research Council, *Productivity and Cyclicality in Semiconductors: Trends, Implications, and Questions,* op. cit.

a faster and cheaper product within the 18-month timeframe, steps up its own work—leading, overall, to the faster pace at which new semiconductor products are brought to market. Upholding Moore's Law, thus, requires keeping up the belief among industry participants that this pace of "faster and cheaper" is sustainable. Continuing this virtuous cycle of expectations requires that each firm in the industry believes that impediments to continuing technological advance can be overcome well in time.

Overcoming Technological Brick Walls

While Moore's Law is currently forecast to hold for the next 10 to 15 years (not least by Gordon Moore himself[76]), there remain potential technological showstoppers down the road. In the case of CMOS (complementary metal-oxide semiconductor) technology, as explained by Bob Doering at the conference on semiconductors, tunneling problems could arise when a gate insulator gets so thin that it loses its insulating capacity and becomes a new leakage path through the transistor.[77] This current flow is dominated by quantum mechanical tunneling of electrons through the barrier.[78]

While Dr. Doering noted that continued advances in CMOS device scaling are expected to continue for another 10 to 15 years, Randall Isaac of IBM, also speaking at the same conference, was more pessimistic, observing that progress from scaling could tail off more rapidly.[79] He noted that the surge in performance, achieved through deep ultraviolet (UV) technologies, is likely not to be sustainable over a long period. He also warned that extreme ultraviolet lithography (EUV), often cited as the next emerging technology, might not prove to be as pervasive as its predecessor has been.

Such technological brick walls apply not only to semiconductors but more broadly to computer components as well. For example, Kenneth Walker, of Philips Electronics, noted at the conference on Deconstructing the Computer that while DVD and CD readers had become standard on personal computers, we are starting to reach certain limits in these devices.[80] Current top-of-the-line CD devices, he noted, rate at 48X to 52X—the equivalent of spinning at about 200 kilometers per

---

[76]*The Economist*, "Moore's Law at 40," op. cit.

[77]CMOS is the semiconductor technology used in the transistors that are manufactured into most of today's computer microchips.

[78]Robert Doering, "Physical Limits of Silicon CMOS and Semiconductor Roadmap Predictions," in National Research Council, *Productivity and Cyclicality in Semiconductors: Trends, Implications, and Questions*, op. cit.

[79]Randall Isaac, "Semiconductor Productivity and Computers," in National Research Council, *Productivity and Cyclicality in Semiconductors: Trends, Implications, and Questions*, op. cit.

[80]Kenneth Walker, "CD/DVD: Readers and Writers," in National Research Council, *Deconstructing the Computer*, op. cit.

hour. This speed approaches the reigning physical limit for CDs, since operating at higher speeds would cause the disc to shred within the device.

Dr. Walker noted that human ingenuity would extend the scope and pace of improvements for hard disks over the near future. The next generation of improvements, he noted, may be realized not by spinning DVDs faster, but by adopting blue lasers to replace red lasers. Since blue lasers are more focused, more information can be stored on a single disk. Newly discovered ways of writing and rewriting information on disks will also enhance the device's functionality, he predicted—although these innovations postpone but do not eliminate a reckoning with the brick wall.

Resource Challenges to Sustaining Moore's Law

In addition to technological impediments, participants at the conference on Productivity and Cyclicality in Semiconductors also reviewed a variety of resource challenges that may jeopardize Moore's Law. These are summarized below.

- ***High Costs of Manufacturing:*** Could the high costs of technical advance be the Achilles heel of the New Economy? Dr. Doering noted that progress on CMOS technology could slow, not because engineers run out of ways to make smaller or faster chips, but because the costs of manufacturing could outstrip the advantages of such miniaturization. Referring to EUV technology, Dr. Isaac noted that at $40 million to $50 million per tool, the economic challenges of investing in such equipment are daunting.

  Dr. Isaac added that the real "fly in the ointment" to computing that is faster and cheaper might well be the cost of power. As engineers place more components closer together, power consumption and heat generation become systemic problems.[81] Though few technologists or economists have factored the cost of power for computing, the energy consumption of server farms is increasing exponentially. To convey a sense of scale, he noted that a server farm uses more watts per square foot than a semiconductor or automobile manufacturing plant.

[81]Randall Isaac, "Semiconductor Productivity and Computers," in National Research Council, *Productivity and Cyclicality in Semiconductors: Trends, Implications, and Questions*, op. cit. Dr. Isaac cited two underlying factors for the accelerating power consumption: The first is that the industry has been following a high-performance scaling law rather than a low-power scaling law. As engineers place more components more closely together, power consumption and heat generation have become systemic issues. The second is that the technology has focused on frequency. A 600 MHz processor uses more than three times the power of a 300 MHz processor. Dr. Isaac suggested that a solution to the power problem might rest with massively parallel systems that use slower but more power-efficient processors.

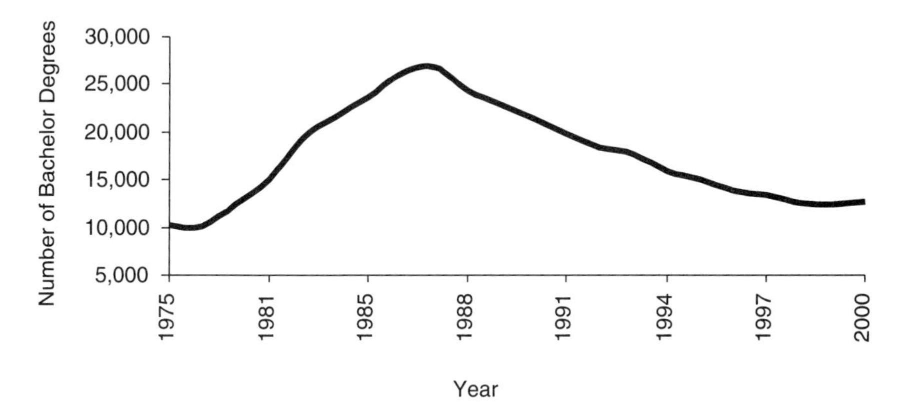

FIGURE 7 Electrical engineering graduates: bachelor's degrees earned, 1975-2000. SOURCE: National Science Foundation, Science and Engineering Indicators 2000, 1975-1987 Engineering Workforce Commission.

- ***Workforce Issues:*** Sustaining Moore's Law will require creativeness and ingenuity in overcoming these technological and economic challenges. George Scalise pointed out that this requires a trained workforce well grounded in the disciplines—such as physics, mathematics, and engineering—that underpin research and manufacturing in the semiconductor industry.[82] Given this need, he listed some recent trends that appear troubling, including:
    - recent evaluations that place American K-12 students below their foreign peers in mathematics and science;[83] and
    - a decline in the number of bachelor's degrees in electrical engineering awarded in the United States (see Figure 7). While this decline (of about 40 percent over the last several years) seems to have recently flattened out, he said that this trend remains a source of concern.[84]

- ***Funding for Research:*** Bob Doering and George Scalise, along with Clark McFadden of Dewey Ballantine LLP, noted that declines in federal R&D funding makes it harder for the semiconductor industry to overcome loom-

[82]George Scalise, "The Industry Perspective on Semiconductors," in National Research Council, *Productivity and Cyclicality in Semiconductors: Trends, Implications, and Questions*, op. cit.

[83]For twelfth grade students in the most recent Third International Mathematics and Science Study (TIMSS), the average score of international students was 500 versus 461 for U.S. students. For additional information on TIMSS, see <*http://nces.ed.gov/timss/*>.

[84]For a detailed discussion of the challenges of maintaining sufficient human capital to sustain the productivity of the semiconductor industry, see National Research Council, *Securing the Future: Regional and National Programs to Support the Semiconductor Industry,* Charles W. Wessner, ed., Washington, D.C.: The National Academies Press, 2003.

ing technological challenges.[85] They added that the semiconductor industry's ability to do its own long-term research has diminished with the demise of the large industrial laboratory.[86] As noted below, several participants called for additional federal investments in research to help maintain the innovative pace of the semiconductor industry.

Strategies to Sustain Moore's Law

Changes in the structure of the semiconductor industry may impact the competitive environment associated with Moore's Law.[87] Kenneth Flamm noted at the conference on Productivity and Cyclicality in Semiconductors that the growth of the foundry model in semiconductor manufacture might have implications for industry-sponsored research, given that foundry-based companies (often called fabs) often spend a smaller percentage of their sales on R&D than do traditional integrated device manufacturers.[88] As George Scalise further noted, fabs have also affected the competitive environment by creating a surge in manufacturing capacity. This surge has led to price attrition beyond levels against which many traditional firms that integrate design and manufacture can compete successfully.

Cooperative Ventures in Semiconductor Research

According to George Scalise and Kenneth Flamm, these developments highlight the importance of sustaining a variety of cooperative efforts to strengthen the

---

[85]National Research Council, *Productivity and Cyclicality in Semiconductors: Trends, Implications, and Questions,* op. cit.

[86]For example, Dr. Doering noted, "Based on physical limits, we need a big R&D effort on many levels to come up with new ideas and take them to a point—even in academic research—where they can be picked up by industry. Where that point of transition between academia and industry is located has shifted today towards academia, because we don't have as many large industrial labs that work at the breadth and depth they used to." Robert Doering, "Physical Limits of Silicon CMOS and Semiconductor Roadmap Predictions," in National Research Council, *Productivity and Cyclicality in Semiconductors*, op. cit.

[87]Dr. Jorgenson asked Dr. Doering at the conference on Productivity and Cyclicality in Semiconductors why the semiconductor roadmap, in predicting product cycles of 3 years, had underestimated the speed at which successive generations of technology are evolving. Dr. Doering responded that the adoption of the two-year cycle was based on "purely competitive factors." See National Research Council, *Productivity and Cyclicality in Semiconductors: Trends, Implications, and Questions*, op. cit., p. 14.

[88]The foundry model separates the electronic design process and the fabrication of physical integrated circuit (IC) devices. In the foundry model, a high-tech company without any semiconductor manufacturing capability (called the fabless company) orders wafer production from a manufacturer (called the merchant foundry.) The fabless design company concentrates solely on the electronic research and development of an IC product, while the foundry concentrates solely on the aspect of fabricating and testing the physical product. See <*http://en.wikipedia.org/wiki/Foundry*>.

research base and to propel advance in semiconductor platform technologies.[89] Positive examples of such cooperative partnerships highlighted at the conference on Productivity and Cyclicality in Semiconductors were:

- ***The Semiconductor Research Corporation (SRC),*** whose mission is to provide low-overhead generic semiconductor research and related programs that meet the needs of the semiconductor industry for technology and relevantly educated talent. It currently disburses approximately $40 million per year on directed research carried out in universities by 800 to 900 graduate students worldwide.

- ***International SEMATECH***, a global research consortium, whose role is to develop new manufacturing technologies and methods and transfer them to its member companies, which in turn manufacture and sell improved chips. Member companies cooperate pre-competitively in key areas of semiconductor technology, sharing expenses and risk. Their common aim is to accelerate development of the advanced manufacturing technologies needed to build future generations of semiconductors.

- ***The Focus Center Research Program,*** which sponsors a multi-university effort to address major basic research challenges.[90] This includes the design and test program led by the University of California at Berkeley, the interconnect team led by the Georgia Institute of Technology, the circuit systems and software team led by Carnegie Mellon University, and a materials and devices team led by the Massachusetts Institute of Technology. Each program has seven to eight partners, and funding for the four-year program, which now totals $22 million a year, is expected to grow to $60 million a year over the next few years.

Expanding the Use of Technology Roadmaps

Technology roadmaps are another important mechanism for sustaining Moore's Law. Providing a graphical portrayal of the structural relationships among science, technology, and applications over a period, a technology roadmap is a tool for firms in an industry to identify potential technical showstoppers and cooperate

[89]See, in particular, remarks by George Scalise, Bob Doering, Kenneth Flamm, and Dale Jorgenson in National Research Council, *Productivity and Cyclicality in Semiconductors: Trends, Implications, and Questions*, op. cit.

[90]The Microelectronics Advanced Research Corporation (MARCO), a cooperative program organized under the auspices of Semiconductor Research Corporation (SRC), funds and operates a number of university-based research centers in microelectronics as part of its Focus Center Research Program (FCRP). For a description of MARCO, see National Research Council, *Securing the Future: Regional and National Programs to Support the Semiconductor Industry*, op. cit.

in developing (at a pre-competitive level) solutions to these technical challenges. Roadmap strategy areas include technology and product marketing, identifying gaps in R&D programs, and identifying obstacles to rapid and low-cost product development. Moreover, as companies believe that competitive success lies in staying ahead of the Roadmap, the existence of a published Roadmap itself enhances the pace of competition and, hence, the robustness of Moore's Law.

At the conference on Productivity and Cyclicality in Semiconductors, Kenneth Flamm noted that although the international semiconductor roadmap is often described as a descriptive or predictive process, its role is to coordinate a complex technology with different pieces and multiple suppliers.[91] "What you really have is people identifying potential showstoppers and trying to mobilize people at choke points." Clark McFadden added that the roadmap is not a "solution" to technological problems but rather a description of various options, challenges, and gaps in charting the future course of a technology. The role of the roadmap, he said, is to communicate information about these options, challenges, and gaps to the industry in a way that suppliers, manufacturers, and customers can appreciate and use.

There are of course limits to the usefulness of roadmaps. As roadmap pioneers William Spencer and T. E. Seidel have acknowledged, roadmaps are expensive and time consuming to develop and are, by definition, out of date as soon as they are written.[92] As they note, however:

> Today, with research and development budgets under pressure in every nation, it's important that redundancy in non-competitive research and development be minimized wherever possible. This is particularly true in major basic research programs in physics, biology, chemistry, and probably computer science. It certainly has been a major help to the U.S. semiconductor industry and the equipment supplier industry for cooperation in pre-competitive technology development.[93]

Crediting the Semiconductor Roadmap for the speed of the information technology industry's recent advance at the conference on Deconstructing the Computer, William Siegle offered two reasons why the road-mapping process is linked to accelerations in the decline of logic cost. First, he noted that making

---

[91]Technology roadmaps are not new to the semiconductor industry. Precursors include a study initiated in the early 1960s by the Committee on Science and Public Policy (COSEPUP) of the National Academy of Sciences. See also G. E. Pakes, *Physics Survey and Outlook,* Washington, D.C.: National Academy Press, 1966; Philip Handler, *Biology and the Future of Man*, London: Oxford University Press, 1970; and D. Alan Bromley, *Physics in Perspective*, Washington D.C.: National Academy Press, 1972.

[92]William J. Spencer and T. E. Seidel, "International Technology Roadmaps: The U.S. Semiconductor Experience," in National Research Council, *Productivity and Cyclicality in Semiconductors: Trends, Implications, and Questions,* op. cit.

[93]Ibid, p. 148.

Box D: Drafting the First Semiconductor Roadmap

Given the complexity of the technology and the multiple participants involved, the need for coordination among the members of the SEMATECH semiconductor consortium arose concerning how best to identify those science and technology areas that have promise and how best to accelerate the transfer of the technology to those useful applications. Following on the footsteps of industry-wide roadmap workshops in June 1987 and March 1988, the Semiconductor Industry Association sponsored a Semiconductor Technology Workshop in 1992, held in Irving, Texas, to develop a comprehensive 15-year roadmap. As recounted by Spencer and Seidel, "The charter of the workshop was to evaluate the likely progress of CMOS technology in key areas relative to expected industry requirements and to identify resources that might best be used to ensure the industry would have the necessary technology for success in competitive world markets."[a]

There were 200 participants at the 1992 workshop, including members of 11 technological working groups assigned to identify issues on specific aspects of semiconductor technology. In preparation for the workshop, these groups developed a "strawman" draft, which was refined through successive review iterations. A revised draft of the roadmap was then issued, with key issues highlighted for review at the actual workshop. The workshop itself included a plenary session, followed by breakout sessions that permitted cross-coordination among the different working groups.

The working format improvised in Texas—"a pretty rushed job compared to how we do it now," as Dr. Doering, an original participant, put it—served as a template for the subsequent 1994 and 1997 roadmaps updates.[b] With the internationalization of SEMATECH, the International Technology Roadmap for Semiconductors (ITRS) was formed in 1998, with a schedule of reports with alternating semi-annual updates and semi-annual full revisions. Under the leadership of the Semiconductor Research Corporation (SRC), the ITRS brings together chipmakers, suppliers, and representatives from SEMATECH and other consortia, along with participants from universities, government, and other relevant organizations to identify future challenges and directions.

---

[a]William J. Spencer and T. E. Seidel, "International Technology Roadmaps: The U.S. Semiconductor Experience," in National Research Council, *Productivity and Cyclicality in Semiconductors: Trends, Implications, and Questions,* Dale W. Jorgenson and Charles W. Wessner, eds., Washington, D.C.: The National Academies Press, 2004, p. 142.

[b]Robert Doering, "Physical Limits of Silicon CMOS and Semiconductor Roadmap Predictions," in National Research Council, *Productivity and Cyclicality in Semiconductors: Trends, Implications, and Questions*, op. cit.

meaningful improvements in capability requires the coordination of many different pieces of technology, and the Semiconductor Roadmap has made very visible both what those pieces are and what advances are required in different sectors of the industry to achieve that coordination. Second, he noted, as companies believe that success lies in staying ahead of the Roadmap, the existence of a published Roadmap enhances the pace of competition.

In these ways, Roadmaps can help sustain the momentum of "faster, better, cheaper" in industries that produce computer components. While welcoming the development of roadmaps for the different computer component industries—such as that recently published by the U.S. Display Consortium[94]—Dale Jorgenson cautioned that successful models, such as the semiconductor industry roadmap, must be adapted to the operational exigencies of the computer component industry in question.

Software and the New Economy

The next conference in the New Economy series examined the importance of software in the New Economy and the vulnerability of the U.S. economy to software failures and attacks. Software is an encapsulation of knowledge in an executable form that allows for its repeated and automatic applications to new inputs.[95] It is the means by which we interact with the hardware underpinning information and communications technologies.

The U.S. economy, today, is highly dependent on software, with businesses, public utilities, and consumers among those integrated within complex software systems. Participants at the NRC Conference on Software, Growth, and the Future of the U.S. Economy, examined how this dependence exposes the economy to vulnerabilities in the production and execution of software—major concerns in sustaining the New Economy.

Almost every aspect of a modern corporation's operations is embodied in software. Anthony Scott of General Motors noted that a company's software embodies a whole corporation's knowledge into business process and methods, adding that "virtually everything we do at General Motors has been reduced in some fashion or another to software."[96]

In addition, much of our public infrastructure relies on the effective operation of software, with this dependency also leading to significant vulnerabilities. As Dr. Raduchel observed, it seems that the failure of one line of code, buried in an energy management system from General Electric, was the initial source

[94]U.S. Display Consortium, "The Global FPD Industry—2003: An In-depth Overview and Roadmap," San Jose, CA.

[95]Monica Lam, "How do we make it?" in National Research Council, *Software, Growth, and the Future of the U.S. Economy,* op. cit.

[96]Anthony Scott, "The Role of Software—What does Software Do?" in National Research Council, *Software, Growth, and the Future of the U.S. Economy*, op. cit.

Box E: Component-Based Software Production

At the conference on Deconstructing the Computer, David McQueeney of IBM recounted the case of a credit card company whose computer system had grown, through a series of ad hoc software patches, so complicated that only three of the company's employees worldwide understood it well enough to manage it when it showed signs of breaking down. He added, however, that added computer complexity is possible if simpler computer architecture makes maintenance easier.

A promising way of addressing this problem of complexity is through component-based software production, which focuses on building large software systems by assembling readily available components. Such components can be used to build both custom enterprise-critical software as well as prepackaged software. Migrating a complex, monolithic system like a credit card system to a newer component-based system, in which updates are handled quickly and efficiently, could lower maintenance costs for firms—and (not least) provide greater security for the nation's financial system by strengthening a critical infrastructure.

However, many of the companies involved in developing component-based software are small start-ups facing severe financing constraints. Because the bulk of their expenditures occur prior to earning any revenues, indeed before technical feasibility has been established, these firms often have difficulty obtaining capital from loans or equity participation. Funding by federal innovation award programs like the Advanced Technology Program (ATP) may be the only way that such technology development projects can be undertaken. Indeed, ATP's focused program in component-based software development is an effort to change the paradigm of custom application to a "buy, don't build" approach for most software projects.[a]

[a]For a recent evaluation of this ATP initiative, see Advanced Technology Program, "Benefits and Costs of ATP Investments in Component-Based Software," NIST GCR 02-834, Gaithersburg, MD: U.S. Department of Commerce, November 2002.

leading to the electrical blackout of August 2003 that paralyzed much of the northeastern and midwestern United States.[97] Smaller, everyday failures are no less expensive; according to the National Institute for Standards and Technology (NIST), national annual costs of software failures lie in the range of $22.2 billion to $59.5 billion.[98]

[97]William J. Raduchel, "The Economics of Software," in National Research Council, *Software, Growth, and the Future of the U.S. Economy*, op. cit.

[98]Based on software developer and user surveys. NIST found that over half of these costs are borne by software users in the form of error avoidance and mitigation activities. The remaining costs are borne by software developers and reflect the additional testing resources that are consumed due to inadequate testing tools. See NIST Planning Report 02-3, "The Economic Impacts of Inadequate Infrastructure for Software Testing," Gaithersburg, MD: U.S. Department of Commerce, May 2002.

Despite the pervasive use of software, and partly because of the relative youth of the science of computer engineering, understanding the economics of software presents an extraordinary challenge. Many of the challenges relate to measurement, econometrics, and industry structure. Here, the rapidly evolving concepts and functions of software as well as its high complexity and context-dependent value make measuring software difficult. This frustrates our understanding of the economics of software—both generally and from the standpoint of action and impact—and impedes both policymaking and the potential for recognizing technical progress in the field.

Given that the infrastructure of the New Economy is based on software, participants at the conference on software considered the vulnerability of this infrastructure and policies that can strengthen this infrastructure.

Making Software More Robust Against Errors and Attacks

Software grows more complex as more and more lines of code accrue to the stack, making software engineering much more difficult than other fields of engineering, according to Monica Lam of Stanford University.[99] This complexity means that the failure of any given piece of software, when added to the stack, may not indicate that something is wrong with that piece of software per se, but quite possibly a failure of some other piece of the stack that is being tested for the first time in conjunction with the new addition. This complexity of software makes it inherently error prone as well as vulnerable to attack.

Indeed, attacks against that code—in the form of both network intrusions and infection attempts—have grown substantially over the past decade, according to Kenneth Walker of Sonic Wall.[100] (See Figure 8.[101]) The perniciousness of the attacks is also on the rise. The Mydoom attack of January 28, 2004, for example, did more than infect individuals' computers producing acute but short-lived inconvenience. It also reset the machine's settings leaving ports and doorways open to future attacks.

The economic impact of such attacks is increasingly significant. According to Kenneth Walker of Sonic Wall, Mydoom and its variants infected up to half a million computers. The direct impact of the worm includes lost productivity owing to workers' inability to access their machines, estimated at between $500 and $1,000 per machine, and the cost of technician time to fix the damage. According to one estimate cited by Mr. Walker, Mydoom's global impact by February 1, 2004,

[99]Monica Lam, "How do we make it?" in National Research Council, *Software, Growth, and the Future of the U.S. Economy,* op. cit.

[100]Kenneth Walker, "Making Software Secure and Reliable, " in National Research Council, *Software, Growth, and the Future of the U.S. Economy,* op. cit.

[101]Figure 8 is based on analysis by Symantec Security Response using data from Symantec, IDC, and ICSA.

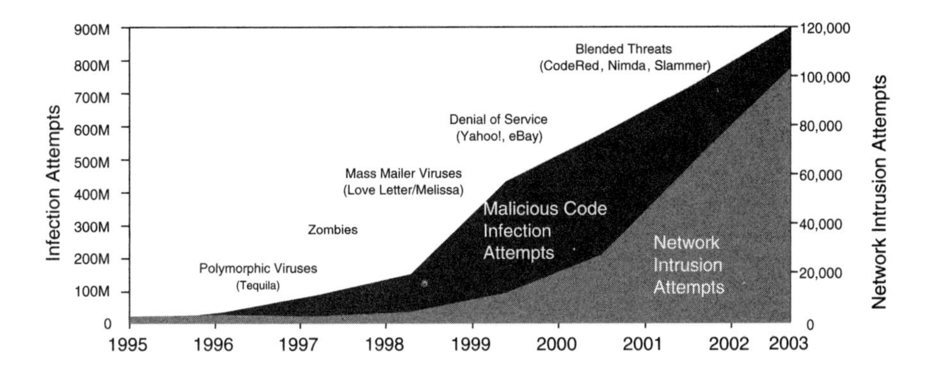

FIGURE 8 Growing attacks against code.
SOURCE: Analysis by Symantec Security Response using data from Symantec, IDC, and ICSA.

alone was $38.5 billion.[102] He added that the *E-Commerce Times* had estimated the global impact of worms and viruses in 2003 to be over one trillion dollars.

Enhancing Software Reliability

Acknowledging that software will never be error free and fully secure from attack or failure, Dr. Lam suggested that the real question is not whether these vulnerabilities can be eliminated, raising instead the issue of the role of incentives facing software makers to develop software that is more reliable.

One factor affecting software reliability is the nature of market demand for software. Some consumers—those in the market for mass client software, for example—may look to snap up the latest product or upgrade and feature add-ons, placing less emphasis on reliability. By contrast, more reliable products can typically be found in markets where consumers are more discerning, such as in the market for servers.

Software reliability is also affected by the relative ease or difficulty in creating and using metrics to gauge quality. Maintaining valid metrics can be highly challenging given the rapidly evolving and technically complex nature of software. In practice, software engineers often rely on measurements of highly indirect surrogates for quality (relating to such variables as teams, people, organizations, processes) as well as crude size measures (such as lines of code and raw defect counts.)

Other factors that can affect software reliability include the current state of liability law and the unexpected and rapid development of a computer hacker cul-

[102]The source of Dr. Walker's estimate is not known. By comparison, BEA reports that all U.S. investment in software for 2004 was $178.5 billion.

ture, which has significantly raised the complexity of software and the threshold of software reliability. While for these and other reasons it is not realistic to expect a 100 percent correct program, Dr. Lam noted that the costs and consequences of this unreliability are often passed on to the consumer.

Addressing this issue, Hal Varian noted that *open-source* software—which, in general, is software whose source code is freely available for use or modification by users and developers—is one way of improving the reliability of software while introducing plural sources of innovation.[103] It is different from proprietary software whose makers do not make the source code available to the public. While developing open-source software provides a public good that is predicted to be under-provisioned in standard economic theory, software developers in the real world have many motivations for writing open-source software, noted Dr. Varian, including (at the margin) scratching a creative itch and demonstrating skill to one's peers. Indeed, while ideology and altruism provide some of the motivation, many firms, including IBM, make major investments in Linux and other open-source projects for solid market reasons.

While the popular idea of a distributed model of open-source development is one where spontaneous contributions from around the world are merged into a functioning product, most successful distributed open-source developments take place within preestablished or highly precedented architectures. It should thus not come as a surprise that open-source has proven to be a significant and successful way of creating robust software. Linux provides a major instance where both a powerful standard and a working reference for implementation have appeared at the same time, noted Dr. Varian. Major companies, including Amazon.com and Google, have chosen Linux as the kernel for their software systems. Based on this kernel, these companies customize software applications to meet their particular business needs.

Indeed, software is most valuable when it can be combined, recombined, and built upon to produce a secure base upon which additional applications can in turn be built. The policy challenge, observed Dr. Varian, lies in ensuring the existence of incentives that sufficiently motivate individuals to develop robust basic software components through open-source coordination, while ensuring that, once they are built, they will be widely available at low cost so that future development is stimulated.

The Software Labor Market and the Offshoring Impetus

Another major and topical issue concerning software and the New Economy concerns the increasingly globalized labor market for software production. Participants at the NRC conference on software discussed the economic forces that

[103]Hal Varian, "Open-source Software," in National Research Council, *Software, Growth, and the Future of U.S. Economy,* op. cit.

are driving this trend, and its implications for sustaining the United States' long-standing advantage in science, research, and innovation.

How is software made and who makes it? Dr. Lam described the software development process as one comprising various iterative stages.[104] After getting an idea of the requirements, software engineers develop the needed architecture and algorithms. Once this high-level design is established, focus shifts to coding and testing the software. She noted that those who can write software at the kernel level are a very limited group, perhaps numbering only in the hundreds worldwide. This division of labor in software production, she said, reflects a larger qualitative difference among software developers, where the very best software developers are orders of magnitude—up to 20 to 100 times—better than the average software developer. This means that a surprisingly small number of people do a disproportionate amount of the field's creative work.[105]

Dr. Raduchel added that as a rule of thumb, producing software calls for a ratio of 1 designer to 10 coders to 100 testers.[106] Configuring, testing, and tuning the software account for 95 to 99 percent of the cost of all software in operation. These non-linear complementarities in the production of software, he said, mean that simply adding workers to one part of the production process is not likely to make a software project finish faster. Further, since a majority of time in developing a software program deals with handling exceptions and in fixing bugs, it is often hard to estimate software development time.

This skew of aptitude in the software labor market means that high-end software firms must look globally to find needed talent, according to Wayne Rosing of Google.[107] Google, he noted, is highly selective. It hired only about 300 new workers in 2003 out of an initial pool of 35,000 resumes submitted from all over the world. While he attributed this high response to Google's reputation as a good place to work, Google in turn looked for applicants with high "raw intelligence," strong computer algorithm skills and engineering skills, and a high degree of self-motivation and self-management needed to fit in with Google's corporate culture.

Google's outstanding problem, Dr. Rosing lamented, was that "there aren't enough good people" available to do this high level of work. Too few qualified computer science graduates were coming out of American schools, he said. While the United States remained one of the world's top areas for computer science

[104]She delineated these stages for analytical clarity, although they are often executed simultaneously in modern commercial software production processes.

[105]Monica Lam, "How do we make it?" in National Research Council, *Software, Growth, and the Future of the U.S. Economy,* op. cit.

[106]This represents Dr. Raduchel's estimate. Estimates vary in the software industry. See William Raduchel, "The Economics of Software," in National Research Council, *Software, Growth, and the Future of the U.S. Economy*, op. cit.

[107]Wayne Rosing, "Hiring Software Talent," in National Research Council, *Software, Growth, and the Future of the U.S. Economy*, op. cit.

education and produced very good graduates, there are not enough people graduating at the master's or doctoral level to satisfy the needs of the U.S. economy, especially for innovative firms such as Google.

In addition, recent U.S. visa restrictions mean that Google must hire the engineers it needs outside the country, said Dr. Rosing. Noting that the government in 2004 capped the H-1B quota at 65,000, down from approximately 225,000 in previous years, he said that Google was not able to hire foreign students who were educated in the United States, but who could not stay on and work for lack of a visa. Dr. Rosing said that such policies limited the growth of companies like Google within the nation's borders—something, he said, that did not seem to make policy sense.

While Dr. Rosing highlighted that the search for talent leads firms like Google to look abroad, Jack Harding of eSilicon noted that manufacturing complexity and business efficiency are often the main drivers of offshore outsourcing.[108] Speaking at the conference on software, Mr. Harding noted that as the manufacturing technology grows more complex, a firm is forced to stay ahead of the efficiency curve through large recapitalization investments or to "step aside and let somebody else do that part of the work." This decision to move from captive production to outsourced production, he said, can then lead to offshore-outsourcing—or "offshoring"—when a company locates a cheaper supplier in another country of same or better quality.

Displaying an outsourcing-offshoring matrix (Figure 9), Mr. Harding noted that it was the actually the "Captive-Offshoring" quadrant, where American firms like Google or Oracle open research and production facilities overseas, that is the locus of a lot of the current "political pushback" about being "un-American" to take jobs abroad. Activity that could be placed in the "Outsource-Offshore" box, meanwhile, was marked by a trade-off where diminished corporate control had to be weighed against very low variable costs with adequate technical expertise.

Saving money by outsourcing production offshore not only provides a compelling business motive, it has rapidly become "best practice" for new companies. Though there might be exceptions to the rule, Mr. Harding noted that a software company seeking venture money in Silicon Valley that did not have a plan to base a development team in India would very likely be disqualified. It would not be seen as competitive if its intention was to hire workers at $125,000 a year in Silicon Valley when comparable workers were available for $25,000 a year in Bangalore. Heeding this logic, almost every software firm has moved or is in the process of moving its development work to locations like India, observed Mr. Harding. The strength of this business logic, he said, made it imperative that policymakers in the United States understand that offshoring is irreversible and learn how to constructively deal with it.

[108]Jack Harding, "Current Trends and Implications: An Industry View," in National Research Council, *Software, Growth, and the Future of the U.S. Economy*, op. cit.

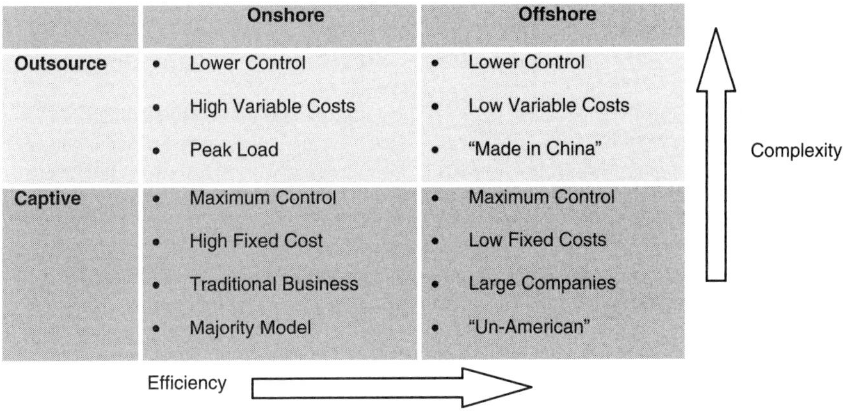

	Onshore	Offshore
Outsource	• Lower Control • High Variable Costs • Peak Load	• Lower Control • Low Variable Costs • "Made in China"
Captive	• Maximum Control • High Fixed Cost • Traditional Business • Majority Model	• Maximum Control • Low Fixed Costs • Large Companies • "Un-American"

Efficiency →

Complexity ↑

FIGURE 9 The offshore outsourcing matrix.

How big is the offshoring phenomenon? Despite much discussion, some of it heated, the scope of the phenomenon remains poorly documented. As Ronil Hira of the Rochester Institute of Technology pointed out at the NRC conference on software, this lack of data means that no one could say with precision how much work had actually moved offshore—a major problem from a policy perspective.[109] Speaking as the chair of the Career Workforce Committee of the Institute of Electrical and Electronics Engineers (IEEE), he noted, nonetheless, that the effects of these shifts were palpable from the viewpoint of U.S. computer hardware engineers and electrical and electronics engineers whose ranks had faced record levels of unemployment in 2003.

Potential Impacts of Offshoring on Future U.S. Innovative Capacities

What is the impact of the offshoring phenomenon on the United States and what policy conclusions can we draw from this assessment? Whereas some economists believe that offshoring will yield lower product and service costs and create new markets abroad fueled by improved local living standards, some leading industrialists have taken the unusual step of arguing that offshoring can erode the United States' technological competitive advantage and have urged constructive policy countermeasures.

Among those with a more macro outlook, noted Dr. Hira, is Catherine Mann of the Institute for International Economics, who has argued that "just as for IT hardware, globally integrated production of IT software and services will reduce

[109] Ronil Hira, "Implications of Offshoring and National Policy," in National Research Council, *Software, Growth, and the Future of the U.S. Economy*, op. cit.

Box F: Two Contrasting Views on Offshore Outsourcing

Outsourcing is just a new way of doing international trade. More things are tradable than were in the past and that's a good thing. . . . I think that outsourcing is a growing phenomenon, but it's something that we should realize is probably a plus for the economy in the long run.

N. Gregory Mankiw[a]

When you look at the software industry, the market share trend of the U.S.-based companies is heading down and the market share of the leading foreign companies is heading up. This x-curve mirrors the development and evolution of so many industries that it would be a miracle if it didn't happen in the same way in the IT service industry. That miracle may not be there.

Andy Grove

[a]Dr. Mankiw made this remark in February 2004, while Chairman of the President's Council of Economic Advisors. Dr. Mankiw drew a chorus of criticism from Congress and quickly backpedaled, although other leading economists supported him. See *The Washington Post*, "Election Campaign Hit More Sour Notes," p. F-02, February 22, 2004.

these prices and make tailoring of business-specific packages affordable, which will promote further diffusion of IT use and transformation throughout the US economy."[110] Cheaper information technologies will lead to wider diffusion of information technologies, she has noted, sustaining productivity enhancement and economic growth.[111] Dr. Mann has acknowledged that some jobs will go abroad as production of software and services moves offshore, but nonetheless holds that broader diffusion of information technologies throughout the economy will lead to an even greater demand for workers with information technology skills.[112]

[110]Catherine Mann, "Globalization of IT Services and White Collar Jobs: The Next Wave of Productivity Growth," *International Economics Policy Briefs,* PB03-11, December 2003.

[111]Lael Brainerd and Robert Litan have further underlined the benefits to the U.S. economy, in this regard, noting that lower inflation and higher productivity, made possible through offshore outsourcing, can allow the Federal Reserve to run a more accommodative monetary policy, "meaning that overall and over time the [U.S.] economy will grow faster, creating the conditions for higher overall employment. See Lael Brainerd and Robert E. Litan, "'Off-shoring' Service Jobs: Bane or Boon and What to Do?" Brookings Institution Policy Brief 132, April 2004.

[112]Challenging the mainstream economics consensus about the benefits of offshore outsourcing, Paul Samuelson has asserted that the assumption that the laws of economics dictate that the U.S. economy will benefit from all forms of international trade is a "popular polemical untruth." See Paul Samuelson, "Why Ricardo and Mill Rebut and Confirm Arguments of Mainstream Economists Supporting Globalization," *Journal of Economic Perspectives,* 18(3), Summer 2004.

Observing that Dr. Mann had based her optimism at that time in part on the unrevised Bureau of Labor Statistics (BLS) occupation projection data, Dr. Hira called for reinterpreting this study in light of the more recent data. He also stated his disagreement with Dr. Mann's contention that lower IT services costs provided the only explanation for either rising demand for IT products or the high demand for IT labor witnessed in the 1990s. He cited as contributing factors the technological paradigm shifts represented by such major developments as the growth of the Internet as well as Object-Oriented Programming and the move from mainframe to client-server architecture.

Dr. Hira also cited a recent study by McKinsey and Company that found, with similar optimism, that offshoring can be a "win-win" proposition for the U.S. and countries like India that are major loci of offshore outsourcing for software and services production.[113] Dr. Hira noted, however, that the McKinsey estimates relied on optimistic estimates that have not held up to recent job market realities. McKinsey's 2003 study found that India gains a net benefit of at least 33 cents from every dollar the United States sends offshore, while the United States achieves a net benefit of at least $1.13 for every dollar spent, although the model apparently assumes that India buys the related products from the United States.

These more sanguine economic scenarios must be balanced against the lessons of modern growth theorists, warned William Bonvillian in his conference presentation.[114] Alluding to Clayton Christiansen's observation of how successful companies tend to swim upstream, pursuing higher-end, higher-margin customers with better technology and better products, Mr. Bonvillian noted that nations can follow a similar path up the value chain.[115] Low-end entry and capability, made possible by outsourcing these functions abroad, he noted, can fuel the desire and capacity of other nations to move to higher-end markets.

Acknowledging that the current lack of data makes it impossible to track activity of many companies engaging in offshore outsourcing with any precision, Mr. Bonvillian noted that a major shift was under way. The types of jobs subject to offshoring are increasingly moving from low-end services—such as call centers, help desks, data entry, accounting, telemarketing, and processing work on insurance claims, credit cards, and home loans—towards higher-technology services such as software and microchip design, business consulting, engineering, architecture, statistical analysis, radiology, and health care where the United States currently enjoys a comparative advantage.

---

[113]McKinsey Global Institute, "Offshoring: Is it a Win-Win game?" San Francisco, 2003.

[114]William Bonvillian, "Offshoring Policy Options," in National Research Council, *Software, Growth, and the Future of the U.S. Economy*, op. cit.

[115]Clayton Christiansen, *The Innovator's Dilemma: When New Technologies Cause Great Firms to Fail*, Cambridge, MA: Harvard Business School Press, 1997.

Another concern associated with the current trend in offshore outsourcing is the future of innovation and manufacturing in the United States. Citing Michael Porter and reflecting on Intel Chairman Andy Grove's concerns, Mr. Bonvillian noted that business leaders look for locations that gather industry-specific resources together in one "cluster."[116] Since there is a tremendous skill set involved in advanced technology, he argued, losing parts of that manufacturing to a foreign country would help develop technology clusters abroad while hampering their ability to thrive in the United States. These effects are already observable in semiconductor manufacturing, he added, where research and development is moving abroad to be close to the locus of manufacturing.[117] This trend in hardware, now followed by software, will erode the United States' comparative advantage in high-technology innovation and manufacture, he concluded.

The impact of these migrations is likely to be amplified: Yielding market leadership in software capability can lead to a loss of U.S. software advantage, which means that foreign nations have the opportunity to leverage their relative strength in software into leadership in sectors such as financial services, health care, and telecom, with potentially adverse impacts on national security and economic growth.

Finally, Mr. Bonvillian pointed out that "manufacturing matters" even in the New Economy. Referring to the work of John Zysman and others, he noted that advanced mechanisms for production and the accompanying jobs are a strategic asset, and their location makes the difference as to whether or not a country is an attractive place to innovate, invest, and manufacture.[118] For the United States, the economic and strategic risks associated with offshoring, noted Mr. Bonvillian, include a loss of in-house expertise and future talent, dependency on other countries on key technologies, and increased vulnerability to political and financial instabilities abroad.

With data scarce and concern "enormous" at the time of this conference, Mr. Bonvillian reminded the audience that political concerns could easily outstrip economic analysis. He added that a multitude of bills introduced in Congress seemed to reflect a move towards a protectionist outlook.[119] After taking the initial step of collecting data, he noted that lawmakers would be obliged to address

---

[116]Michael Porter, "Building the Microeconomic Foundations of Prosperity: Findings from the Business Competitiveness Index," The Global Competitiveness Report 2003-2004, X. Sala-i-Martin, ed., New York, NY: Oxford University Press, 2004.

[117]National Research Council, *Securing the Future: Regional and National Programs to Support the Semiconductor Industry,* op. cit.

[118]Stephen S. Cohen and John Zysman, *Manufacturing Matters: The Myth of the Post-Industrial Economy*, New York, NY: Basic Books, 1988.

[119]Among several bills introduced in Congress in the 2004 election year was that offered by Senators Kennedy and Daschle, which required that companies that sent jobs abroad report how many, where, and why, giving 90 days notice to employees, state social service agencies, and the U.S. Labor Department. Senator John Kerry had also introduced legislation in 2004 requiring call center workers to identify the country they were phoning from.

widespread public concerns on this issue. Near-term responses, he noted, include programs to retrain workers, provide job-loss insurance, make available additional venture financing for innovative start-ups, and undertake a more aggressive trade policy. Longer-term responses, he added, must focus on improving the nation's innovative capacity by investing in science and engineering education and improving the broadband infrastructure.

What is required, in the final analysis, is a constructive policy approach rather than name-calling, noted Dr. Hira. He pointed out that it was important to think through and debate all possible options concerning offshoring rather than tarring some with a "protectionist" or other unacceptable label and "squelching them before they come up for discussion." Progress on better data is needed if such constructive policy approaches are to be pursued.

The Telecommunications Challenge

New telecommunications technologies—the subject of STEP's fifth conference—have contributed significantly to the New Economy. These contributions include the advantages of new product capabilities for businesses and consumers as well as new, more efficient forms of industrial organization made possible by cheaper and more versatile communications. Thus, while the telecom sector accounts, by some measures, for about 1 percent of the U.S. economy, it is estimated to be responsible for generating about 10 percent of the nation's economic growth.[120] A key policy question, therefore, is how to sustain or improve on this multiplier of ten, even as new technological innovations are ushering a major shift from a vertical model to a horizontal model of production and distribution in the communications and entertainment industries.[121] This task of adapting policies and regulations regarding the communications industry to new realities is made more challenging given its long legacy—one that goes back past Alexander Graham Bell to Benjamin Franklin, the first postmaster of the United States.

Communications Technology: A Vision of the Future

Moore's Law, which in its modern interpretation anticipates the doubling of the number of transistors on a chip every 18 months, has spurred the modern revolution in digital technologies for over 40 years.[122] It is likely to continue for

[120]See comments by Dale Jorgenson in National Research Council, *The Telecommunications Challenge: Changing Technologies and Evolving Policies,* Charles W. Wessner, ed., Washington, D.C.: The National Academies Press, 2006.

[121]Dale Jorgenson, "Concluding Remarks," in National Research Council, *The Telecommunications Challenge: Changing Technologies and Evolving Policies,* op. cit.

[122]While by no means dictating an actual law, Moore correctly foresaw in 1965 the rapid doubling of the feature density of a chip, now interpreted as approximately every 18 months. Observing that the number of transistors per square inch on integrated circuits had doubled every year since the

another 10 to 20 years, according to experts in the semiconductor industry.[123] This pace of ever faster and cheaper semiconductors and semiconductor-related technologies is likely to continue to have significant impacts, not least on communications technologies. As William Raduchel noted at the conference on telecommunications and the New Economy, the endurance of Moore's Law means that "the most powerful personal computer that's on your desk today is going to be in your cell phone in twenty years." Technologies for display, storage, and transmission of data are also expected to show rapid improvement, he added, though their rates of improvement are likely to abate sooner than that of semiconductors.[124]

Raduchel predicted that enhanced digital sampling, skyrocketing storage capacity, and expanded packet switching technologies will change the way we will work, communicate, and entertain ourselves in the future.[125] Faster computers mean that digital sampling for recording, playback, looping, and editing of music will improve to the point where it is nearly error free, changing the way music is heard and distributed. Advances in storage capacity and speed will lead to new products (as already previewed with today's iPods and TiVos) that will likely challenge existing business models of how music and video entertainment is packaged and distributed, and ultimately consumed. In addition, advances in packet switching, where information is commoditized for transmission, will likely mean that "radio, television, classified information, piracy, maps, . . . anything" can be moved around a communications infrastructure with no distinction as to what they are. These developments, in turn, will require greater attention to the issue of standards that can allow for coherence as well as future growth and innovation.

These advances in capturing and distributing information and entertainment in commoditized packets build on the concept of the *stupid network*—where the intelligence is taken out of the middle of a communications network and put at the ends—a design principle that has already guided the development of the Internet.[126] According to David Isenberg, such an end-to-end network allows for diversity in the means of transmission—including varieties of wired and wire-

integrated circuit was invented, Gordon Moore predicted in 1965 that this trend would continue for the near future. (See Gordon E. Moore, "Cramming More Components onto Integrated Circuits," op. cit.) The current definition of Moore's Law, which has been acknowledged by Dr. Moore, holds that the data density of a chip will double approximately every 18 months. Many experts expect Moore's Law to hold for another 15 years.

[123]See, for example, Robert Doering, "Physical Limits of Silicon CMOS Semiconductor Roadmap Predictions," in National Research Council, *Productivity and Cyclicality in Semiconductors: Trends, Implications, and Questions*, op. cit.

[124]For a discussion by representatives from these industries of the rate of technological change in these and other computer-related industries, see National Research Council, *Deconstructing the Computer*, op. cit.

[125]See remarks by Dr. Raduchel in National Research Council, *The Telecommunications Challenge: Changing Technologies and Evolving Policies*, op. cit.

[126]David Isenberg, "Rise of the Stupid Network," *Computer Telephony*, pp. 16-26, August 1997.

less technologies—with this diversity creating greater robustness against the failure of any one element. As we see next, enhancements in packet switching capabilities are already making such novel technologies as Voice over Internet Protocols (VoIPs) and Grid Computing technically and commercially feasible for widespread use.[127]

- ***VoIP (Voice over Internet Protocol)***: In Internet telephony, voice is broken into digital packets by a computer and conveyed over the digital network to be reassembled at the other end. The voice network of the future will run over the Internet Protocol, according to Jeff Jaffe of Lucent Technologies. Since this technology has a completely different capability than traditional landlines when it comes to voice quality, cost, and reliability, he predicted that it will bring about a generational change in voice communications.

Louis Mamaokos of Vonage (a company that has introduced VoIP to commercial markets in the United States and elsewhere) cited two sources of opportunity that arise with VoIP: One is through sharing infrastructure, which comes from chopping up audio into packets and transmitting it over an existing packet-based network, which yields significant cost advantages compared with traditional telephony. But equally powerfully, he contended, are opportunities that come from using software to provide a variety of services for the consumer. For example, by marrying it with the computer, phones could be programmed to control who can call through and when.[128]

- ***Grid Computing***: Grid computing, which allows users to share data, software, and computing power over fiber optic networks is expected to be another major development in information and communications technology. Mike Nelson of IBM likens grid computing to a utility supplying electricity, noting that logging onto the Grid could provide a user access to far more computing power than is possible from a single computer system.

A widely known (but limited) instance of the concept of Grid computing is the current SETI (Search for Extraterrestrial Intelligence)@Home project, in which PC users worldwide donate unused processor cycles to help the search for signs of extraterrestrial life by analyzing signals coming from outer space. The project relies on individual users to volunteer to allow the SETI project to harness the unused processing power of the user's computer. About 500,000 people have

---

[127] *The Wall Street Journal*, "Vonage Plans to File for IPO," August 25, 2005.

[128] "For the incumbent telecoms operators, though, what is scary about Vonage is not the company itself but the disruptiveness of its model. Vonage is a telecoms company with the agility of a dotcom. Everyone in the telecoms industry has heard of it, and has wondered what will happen if the model is widely adopted." See *The Economist*, "Between a Rock and a Hard Place," October 9, 2003. We many not have to wait much longer to see what will happen. See *The Financial Times*, "The Internet's Next Big Talking Point: Why VoIP Telephony is Quickly Coming of Age," September 9, 2005, which reports on the entry of Microsoft and Google into the VoIP market.

Box G: VoIP—A Disruptive Technology

VoIP has the potential to undermine the business model underpinning the telecommunications industry. Factors such as the length of the call or the distance between callers, key determinants of cost today, are irrelevant with VoIP. In addition, VoIP augurs more widespread use of videoconferencing as well as new applications such as unified messaging and television over Internet Protocol (IPTV).

Many analysts believe that the question is not whether VoIP will displace traditional telephony, but how quickly. This disruptive potential of VoIP is a challenge for telephone, mobile, and cable incumbents—with some attempting to block the new technology and others moving to embrace it.[a]

[a]*The Economist*, "How the Internet Killed the Phone Business," September 15, 2005. See also Dale Jorgenson, "Information Technology and the World Economy," Leon Kozminsky Academy Distinguished Lecture, May 14, 2004.

downloaded this program, generating an amount of computing power that would have cost $100 million to purchase.

Grid computing is likely to have fewer nodes that are tied together than in the SETI case, said IBM's Nelson, but because the size of the machines can be larger—including large servers, storage systems, and even supercomputers—high levels of computing power can be generated. Further, since the systems involved in Grid computing will be more tightly coupled and more general purpose, they can be far more versatile. The next step in Grid computing, he predicted, is the "Holy Grid" where everything is connected to everything, running common software, able to tackle a wide range of problems. With the advent of such a grid, both small and large companies would be able to buy the computing power they need and get the software they need over this grid of network systems as needed on a pay-as-you-go basis.

In IBM's view, a part of the larger vision of Grid computing includes *autonomic computing*, where integrated computer systems are not only self-protecting, self-optimizing, self-configuring, and self-healing, but also come close to being self-managing. Another important component of this vision is *pervasive computing*, where sensors embedded in a variety of devices and products would gather data for analysis. These sensors will be located all around the world and the data they generate will have to be managed through the Grid. As Nelson predicts, "Soon we will have trillions of sensors, and that is what we really rely on the 'Net for."

The predicted arrival of Grid computing means that firms in the computer industry have an enormous stake in the future of telecommunications networks.

With the Grid, the future of computing lies in complex network-based technologies, such as Web services, which tie together programs running on different computers across the Internet, and utility computing to provide computing power on demand. With telecommunications firms becoming more dependent on information technology, and vice versa, the two industries are likely to become ever more closely intertwined.

While these and other emerging technologies offer alluring prospects for a more vibrant and productive future, a major focus of the STEP conference on telecommunication technologies concerned the regulations that condition the speed at which these technologies and others can be adopted as they become available. As Dr. Jorgenson pointed out in his introductory remarks, the issue of regulation is particularly germane to telecom, which is regulated at both the federal and state levels. Broadband regulation, in particular, was identified by several conference participants as a bottleneck to realizing the benefits of new information and communications technologies in the new "wired" and "wireless" economy.

Sustaining the New Economy: The Broadband Challenge

Broadband, which refers in general to high-speed Internet connectivity, already supports a wide range of applications ranging from email and instant messaging to basic Web browsing and small file transfer, according to Mark Wegleitner of Verizon.[129] In the near future, he said, improved broadband networks can lead to true two-way videoconferencing and gaming as well as VoIP. The future of broadband, he predicted, includes multimedia Web browsing, distance learning, and telemedicine. Beyond these applications, he noted, rests the possibility of immersive gaming and other types of information and entertainment delivery that comes with high band output combined with high-definition receivers.[130]

Can we indeed arrive at this promising future? Charles Ferguson of the Brookings Institution noted that while many foresee what a "radiant future" should look like, there exists an enormous gap for many between this vision for broadband-based technologies and the lack of adequate high-bandwidth access to a broadband network.

[129]Individuals and businesses today variously connect to the nation's fiber-optic network through telephone lines (via digital subscriber lines, or DSL), through television coaxial cables, and by fiber to the home, depending on the availability of these services within different jurisdictions. Wireless connections are also emerging as a viable alternative, as discussed later in the text.

[130]Many of these applications are already emerging, although the potential of many of these applications can be more completely realized through networks that are faster, carry more information, and reach more users.

Indeed, as many conference participants pointed out, the United States is falling behind other nations in access to high-bandwidth broadband.[131] Jaffe drew attention to the reality that the United States had fallen far behind other leading nations in broadband penetration. Isenberg underscored this point, reporting that the International Telecommunications Union (ITU) had, in fact, ranked the United States in thirteenth place in 2003 and that the United States had likely since fallen to fifteenth place in broadband penetration. Citing the ITU figures for 2003, Ferguson reported that the penetration of digital subscriber lines (DSLs) in the United States was 4.8 per 100 telephone lines, in contrast to South Korea where the penetration rate is 27.7 per 100 telephone lines. He noted that the United States had also fallen behind Japan and China in the absolute number of digital subscriber lines.

Acknowledging that this low figure for DSL is explained in part by the fact that a majority of U.S. residential broadband connections are through cable modems, Ferguson nevertheless contended that this fact did little to change the overall picture. In the first place, he explained, when business connections were included, the percentage of total U.S. broadband connections provided by cable was relatively low. In the second place, even in the residential market the percentage of connections provided by cable had been holding roughly constant, as had the cable system's growth rate in respect not only to connections but also to bandwidth levels.

Ferguson observed that bandwidth constraints rather than computer hardware frequently dominate the total cost of adoption of a new network computing application. Personal computers were adequately powerful and relatively inexpensive, he noted, but given bandwidth constraints, deploying a high-performance, high-quality videoconferencing system or other applications could nonetheless prove extremely expensive.

Adding his own negative assessment of the U.S. competitive position, H. Brian Thompson of iTown Communications noted that while (what is commonly called) the Information Superhighway is capable of handling very high capacity in its fiber optic network, and while most desktops and laptops could function at between 1 and 3 gigabits per second, the problem was that there was often less than 1 megabit of connectivity between the two. This weak link—the broadband gap—was illustrated schematically by Thompson at the conference. (See Figure 10.)

[131]Commenting on a discussion of the United States slippage in broadband penetration rates, Dr. Kenneth Flamm of the University of Texas noted that it is important to define carefully what is meant by broadband. Broadband, he noted, describes a wide spectrum of bandwidth, with significant differences between its high and low end. In addition, he noted that while 99 percent of the U.S. population was connected by telephone or cable, and thus was potentially connected to the Internet, the issue of bandwidth size determined the types of applications that could be made practical to households and businesses.

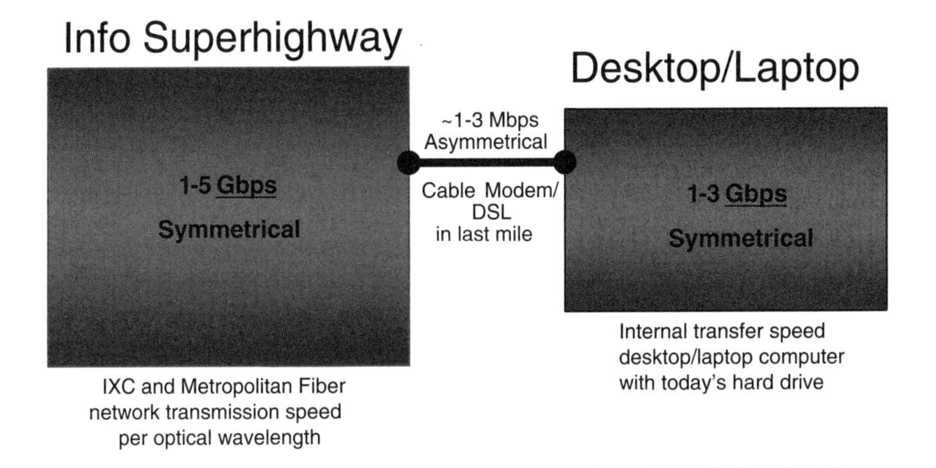

FIGURE 10 The broadband gap: Why aren't current services good enough?
SOURCE: Paul Green, FTTH Council consultant.

Box H: The Demand Side of the Broadband Gap

With much of the discussion on how to address America's apparent lag in broadband adoption focusing on alternative models of service provision, the issue of broadband adoption among users has been relatively obscured. According to the Pew Internet Project's recent survey, the rate of growth in penetration of high-speed Internet at home has slowed and could slow further.[a] While 53 percent of Internet users had high-speed connections at home in May 2005, this level had risen only modestly from 50 percent in December 2004. This is a small and not statistically significant increase, according to Pew's John Horrigan, particularly when compared with growth rates over a comparable timeframe between November 2003 and May 2004 when the adoption rate rose from 35 percent to 42 percent. Horrigan concludes that there is less pent-up demand today for high-speed Internet connections in the population of dial-up users and that this trend is likely to continue. He notes as well that currently 32 percent of the adult U.S. population does not use the Internet at all, and that number is increasingly holding steady.

[a]John B. Horrigan, "Broadband Adoption at Home in the United States: Growing but Slowing," Pew Internet and American Life Project, September 24, 2005. Paper presented to the 33rd Annual Telecommunications Policy Research Conference.

In his remarks at the conference, Mark LaJoie of Time-Warner Cable cautioned that national aggregations showing the United States in thirteenth place worldwide do not tell the whole story. Differences in regulatory climate, the history and condition of infrastructures, the way in which products are used, as well as population densities are all factors influencing measures of broadband penetration. High-density cities like Tokyo and Seoul were likely to have higher levels of penetration, as do similar urban areas in the United States, he said, and added that while the infrastructures in Europe and Asia were newer, U.S. cable and telecom firms were making significant investments in expanding broadband capacity.

Agreeing that there are many ways to spin the numbers on broadband deployment, Mark Wegleitner of Verizon nonetheless acknowledged that "we aren't leading in what we have to perceive as one of the key technologies for any national economic environment going forward." He noted that his company, Verizon, was spending $12 billion annually on improving the broadband infrastructure—including expanding fiber to the home—thereby helping the United States catch up with other leading nations. At the same time, he predicted that "bandwidth demands are just going to grow and grow and grow," as new applications come into use.

Implications of the Broadband Gap

If broadband can serve as an engine for the nation's future growth and competitiveness, as emphasized by several participants at the conference, a lack of an adequate access to the broadband network may lead to a loss of this economic opportunity.[132] Assessing the impact of the broadband gap, Charles Ferguson noted that the "local bandwidth bottleneck" is having a substantial negative effect on the growth of the computer industry and of various other portions of the information technology hardware and software sectors. While conceding that computing an estimate of this impact in a rigorous way would be extremely difficult, he nevertheless asserted that "you can convince yourself easily that this effect is something on the order of one-half of one percent—or even up to one percent—per year in lost productivity growth and GNP [Gross National Product]."

Commenting on the national security implications of the broadband gap, Jeff Jaffe reminded the audience that the 9/11 Commission had recommended that the nation's digital infrastructure be prepared to deal with simultaneous physical and cyber attacks. In the case of a national emergency it will be important for first responders and other individuals to communicate effectively with each other, and

[132]Dr. Raduchel, for example, noted that new technologies like embedded sensors, which rely on a capable broadband network, could emerge as the source of the next round of productivity improvements.

a high-bandwidth, interoperable system is essential for this task, he said, adding that such a network is still not in place today.

Some Explanations for the Broadband Gap

While many of the participants at the conference concurred that the United States faces a broadband gap, views varied as to the reasons for as well as solutions to this situation. Some suggested that the broadband gap has emerged because some telecom and cable companies have been reluctant to provide adequate interface between the user and the fiber optic cable networks. Others suggested that the broadband gap arose from the consequences of federal and state regulations.

- ***Flawed Market Motives of Telecom and Cable Companies***: What is holding back high-bandwidth broadband penetration in the United States? Dr. Isenberg noted that the rise of the stupid network makes it difficult for the telephone or fiber company to sell anything other than commodity connectivity. In the new inter-networked model, it was the Internet Protocol's job to make all that was specific to a single network disappear and to permit only those things common to all networks to come to the surface. Since the Internet ignores whatever is specific about a single network, including features that had formed the basis of competition for the telephone or cable companies, these companies have little to sell beyond access, he argued, and therefore faced little incentive in providing the public access to high-bandwidth broadband. The result, he said, was a crippled network with far less bandwidth available than technology would allow or than is available in other technologically advanced countries.

Ferguson suggested that flawed markets were behind the high cost of securing adequate bandwidth in the United States. He noted that both the telephone and the cable companies had "severe conflicts of interests" and that they largely avoided competing with each other. Even competition for residential markets was "quite restrained, and much less substantial than you might suspect."

The conflict of interest for the telephone companies is "fairly obvious," Ferguson asserted. Incumbent businesses were providing very expensive voice and traditional data services. Very rapid improvements in price/performance of bandwidth would undercut their dominant businesses in a major way. The same was true of the cable system: It provided video services that could easily be provided over a sufficiently high-performance Internet Protocol network.

- ***Consequences of Unbundling Network Elements***: In the discussion following the second panel, Kenneth Flamm noted that more than one speaker had spoken of a tendency to dismantle some of the opening up of the local loop that had been the centerpiece of the 1996 Telecommunications Reform Act. The Act required incumbents to make parts of its network available to competing operators, in particular the "local loops"—the wires that run from telephone exchanges into homes and offices.[133]

The 1996 Act sought to promote competition by asking incumbents to share this part of their networks with rivals—technically known as "local loop unbundling" (LLU)—given that the expense for competitors to build their own networks would be very high in the short term. In practice, however, most incumbent operators saw unbundling as robbery, according to Thompson. This meant (as *The Economist* describes it) that "the incumbent must, in effect, give its rivals a hand as they try to steal its business. Not surprisingly, most incumbents find procedural, legal and technical reasons for being slow about it."[134] Though intended to promote competition in the short run, local loop unbundling may have inhibited investments in alternate infrastructure that competitors might otherwise have made over the longer term. And because it forced incumbents to share their networks with rivals, this may have also deterred them from investing in new equipment. An unintended consequence of the 1996 Telecommunications Act may well have been to inhibit investment needed to provide high-bandwidth broadband access over the local loop, although the issue of whether mandatory unbundling increases or decreases the roll out of broadband network access remains an open empirical question.

Even so, one of the authors of the Telecommunications Act of 1996, Charles Thompson, conceded that the concept of unbundled network elements, introduced in that legislation was moribund—that he "would be the first to put flowers on the grave of unbundled network elements."

- ***Outdated Standards and Regulatory Uncertainty***: Outdated standards and a regulatory uncertainty may be retarding progress in addressing the broadband gap, according to some conference presenters. On the issue of standards, Peter Tenhula of the Federal Communications Commission (FCC) acknowledged that wireless technology regulation was still being governed by a 90-year-old spectrum management regime rather than one "rooted in modern-day technologies and markets." Such outdated regulations, he noted, fail to capitalize on technological advances in digital technologies such as those that allow for greater throughput of information, interference management, and spectrum sharing.

Regulatory uncertainty is also holding down the installation of fiber all the way to the curb, noted Dr. Jaffe. Clear regulation is needed, he stated, to

[133]Local loops can be either "legacy" copper loops or newer fiber broadband connections. The 1996 Telecom Act created considerable uncertainty for the unbundling broadband services. See, for example, the press release of April 8, 2002, by the Telecommunications Industry Association, "TIA Tells FCC That Unbundling Rules Discourage Broadband Investment," which recommends that the FCC (Federal Communications Commission) not apply its network unbundling rules to new facilities used for the provision of broadband and high-speed Internet access services, and to apply them to legacy systems including copper loops, so as not to inhibit investment in wire-line broadband networks.

[134]*The Economist*, "Untangling the Local Loop," October 9, 2003.

encourage sufficient near-term investment in fiber infrastructure. This regulatory environment may have been further clouded in recent years by increasing federal concerns about infrastructure protection, disaster recovery, and emergency services in the wake of recent concerns about terrorism. According to Jaffe, vendors such as Lucent face uncertainties in developing new products at a time when regulatory imperatives are very slow to come out.

Another important source of regulatory uncertainty is the patchwork of local regulations issued by individual municipalities. Cable infrastructure is often governed by city-specific franchise agreements, while telephone companies and other broadband providers may in some cases prefer statewide or even national authority as a means towards greater regulatory simplicity and predictability.

In addition, as Verizon's Wegleitner observed, prevailing uncertainties in updating regulation make it difficult for his company to invest in the development of an effective broadband network. Incremental rulemaking in the transition from the old regulatory regime to a new one often creates ambiguities, with investments of millions or even tens of millions of dollars hinging on the interpretation of words that, while written only a few years before, were already technically obsolete. "It is that interpretation that is going to determine the path forward of the network's evolution." This "unnecessarily complex regulatory environment" did not make sense in that it discouraged investment.

Thompson objected, however, arguing that large telecom and cable companies are not passive recipients of federal and state regulation and that, moreover, the current regulatory environment is greatly affected over the years by the power of incumbents on all sides. To the extent that incumbents influence regulation, the current uncertainty in regulation may well reflect the uncertainties that major cable and telecom providers are facing in coming up with a viable business model that allows profits in an arena that has been transformed by new technologies. Lisa Hook, recently of AOL-Broadband, noted in this respect that firms in the broadband industry were struggling at the service layer to find business models and revenue streams based on new technologies that would justify the investment needed to make nearly unlimited bandwidth widely available.

Some Alternative Solutions to Close the Broadband Gap

According to IBM's Nelson, the Internet revolution is less than 8 percent complete, with many new applications still to be enabled by future technologies like the Grid. Realizing this vision of the next-generation Internet will require both new technologies as well as significant investment, he cautioned, as it will entail providing whole neighborhoods with gigabits-per-second networks that are as affordable and reliable as they are ubiquitous. "Getting there is going to require more intelligent, more consistent policies than we have today," he declared. Participants at the conference considered a variety of means by which the nation could close the broadband gap, of which some key approaches are previewed below.

- ***Directed Government Incentives***: Ferguson suggested that the nations that were ahead of the United States in broadband penetration shared two characteristics. The first was that their governments are "much more heavily involved in providing incentives and/or money and/or direct construction of networks than is the case in the United States." The second was that their Internet providers are under government pressure to improve their price and performance. For example, he said that the Chinese government had made it clear to the country's principal telecommunications providers that broadband deployment was a major national priority. The situation was similar in Japan and Korea, adding that government encouragement in Canada and the Scandinavian countries had also enabled those countries to surge ahead of the United States in high-bandwidth broadband penetration.[135]

For the United States, Ferguson recommended a variety of policy measures to bridge the broadband gap. Initiatives could include subsidizing the deployment of municipal networks and offering investment incentives to public and private providers. Putting more pressure on incumbents to open up their networks so that there is an open architecture broadband system that is more analogous to the structure of the Internet is another avenue.

- ***Faith in Efficient Markets***: In contrast to this more policy-driven approach, Verizon's Wegleitner noted that broader technical, financial, and regulatory improvements would reduce uncertainty and allow markets to function efficiently. While admitting that current challenges resisted simple solutions, he put forward what he called a short answer to the problem: "Let the markets rule." By this, he envisioned the Internet of the future as an *interconnection of commercial networks* such as Verizon's rather than the *confederation of commercial providers* that it is now. He added that the future requirements for services offered customers via broadband would be of such quality and scope that only an interconnection of commercial networks could provide this service.[136] To make this network of the future possible, Wegleitner recommended further development of appropriate standards for communication protocols and a new way of levying tolls on customers for use of the infrastructure that belongs to companies like Verizon, combined with a light regulatory touch.[137]

---

[135]For an assessment of Japanese policies to catch up and surpass the United States in broadband connectivity, see Thomas Bleha, "Down to the Wire," *Foreign Affairs*, 84(3):0015-7120, 2005.

[136]The current Internet is based on a confederation made up of multiple service providers. Their ability (or inability) to maintain their interconnection arises from commercial issues, and not from the current design of the Internet.

[137]Responding to such proposals by the telcos, proponents of "net neutrality" have argued that basic Internet protocols should remain neutral with respect to the diverse ways in which they can be used.

- ***Networks in the Hands of Customers***: In the discussion that followed the first panel, Jay Hellman, a real estate developer, observed that there exist business opportunities both in laying fiber to the home and in making sure it functions. He likened the duo of fiber and services to a public roadway where service companies like FedEx and UPS competitively ply their fleets. It was desirable, he added, that the street be accessible to as many competitors as possible. He also added that his own frustration with the capacity offered by existing providers had prompted him to start his own small telecommunications company. Responding to this comment, David Isenberg noted that the development of technologies that allow customers to create their own networks and that create opportunities for individuals to provide service innovations was important to sustain innovation and provided a broader, more generic solution to the broadband challenge.

- ***Municipally Owned Fiber***: Thompson proposed a different approach, recommending the development of non-profit public-private partnerships at the local level to stimulate the development of broadband to the home. These partnerships would serve as a utility, lighting fiber but not providing any service on that fiber except those municipal services that the town or community base chose to provide. The network would be open to any and all service providers with an Internet Protocol basis—be they telephone companies, cable companies, software companies, or others providing on-line entertainment—and it would be used by all under the same terms and prices. Communities could build this network, just as municipalities build and maintain roads and sewers, he added, citing the case of Ireland where, Thompson said, such partnerships have been successfully developed to provide broadband access.

While separating the network access component from retail services may help municipal providers of network infrastructure, more needs to be learned about the feasibility of this idea in the United States, including whether customers want to buy their services in this way. The issue of whether the municipal provision of infrastructure will in fact lead to more competition for broadband access also remains to be studied.

- ***The Wireless Wildcard—A Silver Bullet?***

Wireless broadband access can be a third tier that competes with cable and DSL, according to David Lippke of HighSpeed America.[138] In this way, wireless broadband can help overcome the limitations associated with traditional wired

---

They argue that net neutrality protection is critical for the Internet to continue to meet its innovative promise. Others argue that recouping all new network construction costs from consumers alone could drive up prices or discourage investment, exacerbating the broadband gap. See *The Wall Street Journal*, "'Net Neutrality' Snags Overhaul of Telecom Laws," June 29, 2006, p. A7.

[138]Also mentioned at the conference was broadband over power lines, which at the time was being reviewed by the FCC.

broadband access. While wireless broadband has been in limited use so far due to relatively high subscriber costs and technological limitations such as problems with obstacle penetration, rapid advances in technology are likely to overcome such challenges. Moore's Law applies to wireless no less than other forms of telecommunications, he noted, predicting that wireless data rates would reach all the points through which traditional telecom had passed.

In particular, scientists and engineers working on the upcoming WiMAX standard have resolved a number of problems that had bedeviled existing wireless protocols such as WiFi. The prospect of reaching gigabit speeds was now being mentioned, and other quality-of-service issues as well as lower costs of installation are being addressed. To the extent that these predictions are realized, the WiMAX protocol may well offer an effective wireless solution to the broadband gap, especially for smaller towns and communities across the United States.

The End of Stovepiping

The move from analog to digital information and communication technologies is ushering a major transformation disrupting how telecom, cable, and music and video entertainment companies, among others, do business. Because analog solutions were all that existed until recently (except in some fields of computing), these industries each matured into separate industries, with separately evolved business models and regulatory frameworks. In the digital age, however, basic technologies like digital sampling and packet switching enable the commoditization of voice, data, and images into digital packets that resemble each other. These packets can be sent over the Internet with no distinction as to what they are, to be reassembled at the intelligent ends of the network.

Drawing on these observations, William Raduchel noted at the conference that the information and communications technology revolution will usher the end to stovepiping as service and content providers shift from vertical integration to a greater reliance on horizontal platforms. This change, he noted, will give rise to a variety of public policy issues as individuals and businesses in the economy restructure to take advantage of the potential offered by new technologies.[139] He also noted that the speed of change is likely to be such that the economy may not be able to adjust to it readily. Among the issues to be addressed is the challenge to intellectual property rights and the question of regulation, which is expected to be very challenging.

The potential and implications of the move from analog to digital information and communication technologies were discussed by several of the conference's participants. Key points from these discussions are summarized below. As in any

[139]A key example of contemporary relevance is the offshore outsourcing issue. For a discussion of this issue, see National Research Council, *Software, Growth, and the Future of the U.S. Economy*, op. cit. See also Catherine L. Mann, *High-Technology and the Globalization of America*, forthcoming.

Box I: Some Factors Affecting the End of Stovepiping

While the digital transformation has the potential to disrupt traditional vertically integrated industrial organizations, some factors may inhibit a transformation to a fully horizontal platform.

- ***Open Network Architecture***: The horizontal organization of communications requires a relatively open network architecture. However, if systems or content providers do not have access to physical or logical pipes, those providers cannot reach their customers.[a]
- ***Separation of Carriage from Content***: Some customers may prefer to purchase services in bundles that include access, as noted by Lisa Hook. Here, vertically integrated firms may have a competitive advantage over firms that supply pipes or content exclusively.
- ***Social Policies that Favor Universal Access***: Where social policies set access price below a competitive market price, the supplier of the access must also be able to cover its total cost from the supply of some other higher-margin services or receive a subsidy.
- ***Economies of Scope***: There may be economies of scope between providing communications services and network facilities.

[a]Consider, for example, the FCC's Video Dialtone initiative in the 1990s, which attracted substantial investment from incumbent telephone companies until it was determined that some portion of the bandwidth had to be made available to competing content providers. For a wider discussion of the limitations of open access cable, see Thomas W. Hazlett and George Bittlingmayer, "The Political Economy of Cable 'Open Access,'" *Stanford Technology Law Review*, 4, 2003.

conference that includes a variety of perspectives, some of these policy recommendations are mutually contradictory, and evidence may be required regarding their efficacy.

Convergence and Competition

Raduchel sees the Internet as having two complementary aspects—it is both a physical set of networks as well as a protocol known as TCP/IP. At present, the physical network can only support movies and other applications at low bit volumes and is often not cost-effective—although this can be expected to change as technology improves and the broadband gap is overcome. The significance of the Internet Protocol, he said, is that it makes all networks look the same and allows interoperability. It was for this reason that the telecommunications world could be expected to move to one set of interconnected webs, he said, predicting that "five to ten years from now, we will be online all the time."

This convergence is challenging the traditional business models of firms in these industries. How would telecom companies, for example, deal with new technology that makes cell phones work perfectly everywhere or with much cheaper VoIP service? The next decade, warned Dr. Raduchel, would be marked by "lots of dislocation" as firms attempt to adjust to new technological and commercial realities.

According to Mr. LaJoie, the convergence of data, voice, video, wireless, public networks, and private networks in an end-to-end infrastructure was changing the terms of competition across industries. Where there was once a big separation between what the telecom and cable industries did for example, "now everybody is in everybody else's business." While cable television, Internet, Cellular, WiFi, and satellite transmission businesses were once distinct, LaJoie believes that they are all destined to overlap and offer similar kinds of products, suggesting with some optimism that the economic rewards that will arise from this competition would be what drives continued innovation, the advent of new services, and increased broadband connectivity.

The potential end of stovepiping also poses new challenges for consumers. Many consumers, faced with a proliferation of Internet services, operating systems, and devices will want a service that is easy to use and integrated, predicted Ms. Hook. She noted that companies like AOL Broadband see a market opportunity as aggregators, packaging a variety of content and communications services over the Internet and protection against viruses and spy-ware that are easy to launch and use.

Intellectual Property in the Era of Digital Distribution

In addition to disruption in the business models of firms that deliver a digital signal is the disruption to business models of firms that provide the content. Indeed, the music and entertainment industries are among those that are also undergoing a fundamental shift in the digital age. Andrew Schuon of the International Music Feed television network noted that while the public's desire to consume music has never been greater, with new technologies allowing users to take an entire music collection with them anywhere they go, the key problem for content providers is how to make money selling music in the new medium—given that technology already available has allowed consumers to share music and other content with each other for free. At present, he noted, legitimate downloads account for only a few percent of all downloads from the Internet.

He noted that technology developed for building legitimate services makes it now possible to protect intellectual property, to monetize it, and to track licenses while, at the same time, creating a good experience for the consumer. However, this technology has to catch up with consumer expectations that have developed in the absence of such constraints: "If you steal the content, you can do anything you want with it—put it into any portable device, put it on as many computers

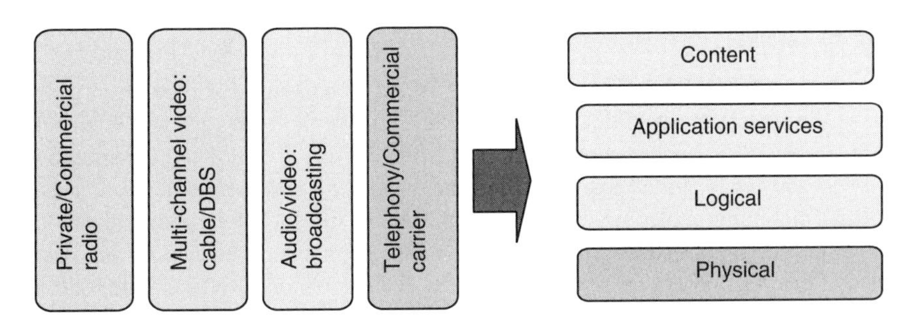

FIGURE 11 Vertical silos to horizontal layers.

as you have, use the content as you see fit." The challenge for the music industry is to find a way to get the consumer to pay for its product while at the same time being more creative than the illegitimate download sites. The music industry, Mr. Shuon said, has to offer the modern customer the flexibility to use the content in the way they want to, in addition to offering superior content and a fair price.

Steve Metalitz, of the law firm Smith and Metalitz, agreed that developing a legitimate market for copyrighted materials over broadband—for entertainment, services, software, video games, research and reference works—was indispensable for the long-term viability of these industries. Acknowledging that piracy will continue to be a problem, he added that the challenge for the future of broadband is to achieve a relatively low level of piracy and a very high level of legitimate products. Addressing this challenge requires:

- developing legitimate markets for copyrighted materials over broadband,
- providing greater security for delivering content to an end-user including measures to ensure that the income-generating potential of material going into the pipe did not vanish forever,
- creating a usable legal framework to protect the technological measures used to control access to copyrighted material in the network environment,
- focusing enforcement of piracy problems on organized criminal groups as well as dedicated amateurs who play a role in making the system insecure, and
- improving public education to make consumers aware that certain types of file sharing are illegal and of the need to secure permission to avoid copyright infringements.

Cooperation, Mr. Metalitz concluded, is needed among providers of network services along with better communication with policymakers to advance these objectives.

The Challenge for Regulation

According to Peter Tenhula of the FCC, the challenge for regulation concerns the migration from decades of regulatory stovepipes towards a new vision of a variety of applications and services (covering voice, video, and data among others) that are provided over multiple and competing telecommunications platforms (including cable, satellite, DSL, and power lines). For this idea to work, content or service providers need a choice of mechanisms by which they can reach their customers. Rather than preserve the artificial vertical integration that had existed for decades and had created silos that grew up over the years, Mr. Tenhula suggested that it made better sense to let the natural layers fall as they might. (See Figure 11.) Replacing sector-specific communications regulation with a layered regulatory model, he added, would better complement the networked characteristic of the New Economy.

The FCC's agenda, he said, was to guide and propel the journey from a slow, conventional analog world to a digital world with significant opportunities for faster, more reliable, higher-quality information and communications, with the overall goal of providing substantial benefits for American consumers.

Towards a New Agenda of Research

Concluding the series of conferences on the New Economy, Dr. Jorgenson noted that the New Economy had witnessed a huge shift from a vertical model to a horizontal model in the computer, semiconductor, and communications industries. In this new model, he said, most of the interesting innovations were disruptive. The challenge for businesses in this changing environment was to figure out how to make money, which was hard given that consumers were both clever and unpredictable. It was "too bad," he said, that the consumer ends up carrying away most of the welfare, which then cannot be delivered to shareholders. But in another respect, he added, the fact that "consumers emerge over and over again as the big winners . . . was a great thing about the New Economy."

Dr. Jorgenson characterized the policy issues in the telecommunications challenge as particularly difficult. While many economists are prone to offer private property as an answer to policy dilemmas, the presence of common property in the form of the digital communications infrastructure made matters more complex, he noted, adding that a way had to be found of maintaining common facilities within a market-based approach. The transmission of property such as data, software, and music across this network also raised questions about its protection, while ensuring privacy for users. Taken together, these issues provide a robust agenda for further study and consideration about the New Economy—which, he noted, has been a central aim of the National Academies' Board on Science, Technology, and Economic Policy.

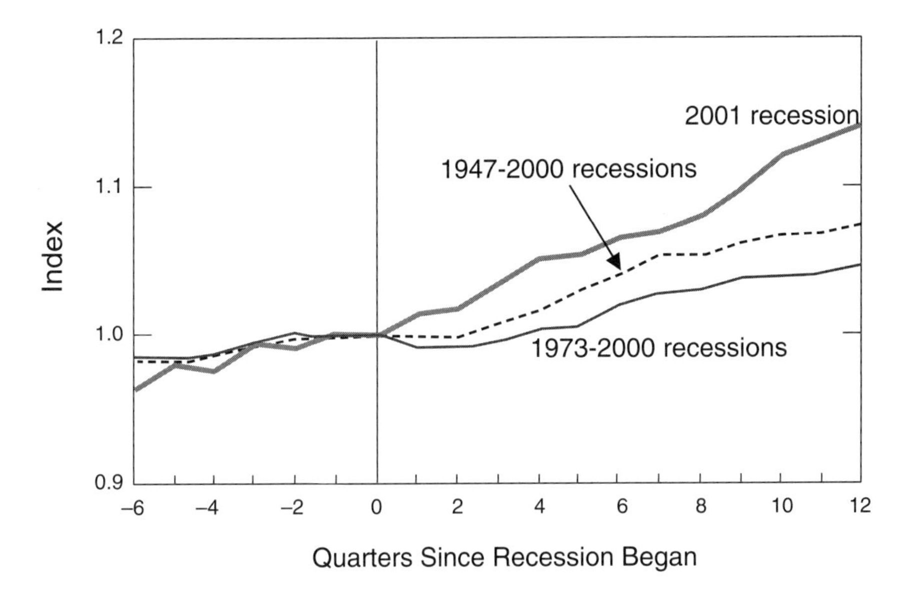

FIGURE 12 Productivity growth over the business cycle: 2001 recession compared with averages of earlier recessions.
SOURCE: U.S. Department of Labor, Bureau of Labor Statistics.
NOTES: Productivity series are normalized to equal 1.0 at the beginning of each recession. The 1973-2000 line represents average productivity growth over the four recessions during that period; the 1947-2000 line represents average productivity growth over the nine recessions during that period.

THE FUTURE OF THE NEW ECONOMY

The New Economy is alive and well today. Recent figures indicate that since the end of the previous recession in 2001, productivity growth had been running about two-tenths of a percentage point higher than in any recovery of the post-World War II period.[140] (See Figure 12.) The challenge rests in developing evidence-based policies that will enable us to continue to enjoy the fruits of higher productivity in the future. It is with this aim that the Board on Science, Technology, and Economic Policy of the National Academies has undertaken a series of conferences to address the need to measure the parameters of the New Economy as an input to better policymaking, and to highlight the policy challenges and opportunities that this New Economy offers.

[140]Dale W. Jorgenson, Mun S. Ho, and Kevin J. Stiroh, "Will the U.S. Productivity Resurgence Continue?" *Current Issues in Economics and Finance,* 10(13), November 2004.

This ambitious series was begun in the midst of a tremendous economic boom, and although economic conditions have changed since then, the basic structural dynamics underpinning the New Economy have remained intact. Faster and cheaper computing power and communications capabilities continue to have a momentous impact on productivity growth in the United States and around the world. Understanding the basis and dimensions of this New Economy is important if we are to develop the economic policies required to ensure the nation's future prosperity and growth.

> "STEP has produced the most detailed and comprehensive picture of the New Economy available to date."
>
> —Dale Jorgenson

STEP's series of conferences on the New Economy have given momentum to this task. STEP's first conference on Measuring and Sustaining the New Economy showed that technology is the main source of the development denoted by the term "New Economy," and that the key technologies center on semiconductors.

The second conference addressed semiconductors specifically, dealing—as described by Moore's Law—with the speed at which semiconductor technology develops. At that conference, Robert Doering of Texas Instruments and other leading authorities in the field projected that semiconductor development would continue at that accelerated pace for at least another decade or so while highlighting what needs to done to keep Moore's Law on track.

The topic of the third conference in the series was computers. That conference brought to light that the industries that manufacture computers and computer components are also driven by a Moore's Law phenomenon and that they have developed internal metrics to gauge rapid technological developments.

The fourth conference of the series examined developments in software measurement, the vulnerabilities affecting the nations' complex software infrastructure, as well as implications of the offshoring of software production abroad.

The final meeting on the telecommunications challenge described a huge shift from a vertical model to a horizontal model of production made possible by inexpensive computing and communications. Low-cost and rapid data and voice transmission is transforming the competitive strengths of national economies by ushering the rapid globalization of research and production. How we adapt our laws and regulations to capitalize on these new technological opportunities will determine the future of the United States' security and economic preeminence in the world.

Taken together, the work sponsored by STEP under the rubric Measuring and Sustaining the New Economy has produced what Dale Jorgenson described as the most detailed and comprehensive picture available to date of what is known as the New Economy. This undertaking provides the basis for further research on the dimensions of the New Economy and policies that can enhance the benefits of the New Economy.

IV

BIBLIOGRAPHY

Bibliography

Abel, Jaison R., Ernst R. Berndt, Cory W. Monroe, and Alan White. 2004. "Hedonic Price Indexes for Operating Systems and Productivity Suite PC Software." Draft Working Paper.

Abel, Jaison R., Ernst R. Berndt, Alan G. White. 2003. *Price Indexes for Microsoft's Personal Computer Software Products*. NBER Working Paper 9966. Cambridge, MA: National Bureau of Economic Research.

Abramovitz, Moses, and Paul David. 1999. "American Macroeconomic Growth in the Era of Knowledge-Based Progress: The Long Run Perspective." In *Cambridge Economic History of the United States*. Robert E. Gallman and Stanley I. Engerman, eds. Cambridge, UK: Cambridge University Press.

Advanced Technology Program. 2002. "Benefits and Costs of ATP Investments in Component-Based Software." NIST GCR 02-834. Gaithersburg, MD: U.S. Department of Commerce.

Aizcorbe, Ana. 2005. "Moore's Law, Competition, and Intel's Productivity in the Mid-1990s." BEA Working Paper WP2005-8. Washington, D.C.: Bureau of Economic Research.

Aizcorbe, Ana. 2005. "Why are Semiconductor Price Indexes Falling so Fast? Industry Estimates and Implications for Productivity Measurement." BEA Working Paper WP2005-7. Washington, D.C.: Bureau of Economic Research.

Aizcorbe, Ana, Kenneth Flamm, and Anjum Khurshid. 2002. "The Role of Semiconductor Inputs in IT Hardware Price Decline: Computers vs. Communications." Federal Reserve Board Finance and Economics Series Discussion Paper 2002-37. Washington, D.C.: Board of Governors of the Federal Reserve System.

Aizcorbe, Ana, Kenneth Flamm, and Anjum Khurshid. Forthcoming. "The Role of Semiconductor Inputs in IT Hardware Price Declines." In *Hard to Measure Goods and Services: Essays in Honor of Zvi Griliches*, E. Berndt, ed. Chicago, IL: National Bureau of Economic Research.

Aizcorbe, Ana, Stephen D. Oliner, and Daniel E. Sichel. 2003. "Trends in Semiconductor Prices: Breaks and Explanations." Washington D.C.: Board of Governors of the Federal Reserve System.

Anderson, Lenda Jo, et al. 1995. "Discovering the Process of Mass Customization: A Paradigm Shift for Competitive Manufacturing." National Textile Center Annual Report.

Aron, Debra J., Ken Dunmore, and Frank Pampush. 1998. "Worldwide Wait? How the Telecom Act's Unbundling Requirements Slow the Development of the Network Infrastructure." *Industrial & Corporate Change,* 7(4):615–621.

Association of Computing Machinery. 2006. "Globalization and Offshoring of Software." William Aspray, Frank Mayadas, and Moshe Y. Vardi, eds. New York, NY.

Baily, Martin N. 2002. "The New Economy: Post Mortem or Second Wind?" *Journal of Economic Perspectives* 16(2):3–22.

Baily, Martin N. and Robert J. Gordon. 1998. "The Productivity Slowdown, Measurement Issues, and the Explosion of Computer Power." *Brookings Papers on Economic Activity* 2:347–420.

Baily, M. N. and R. Z. Lawrence. 2001. "Do We Have an E-conomy?" NBER Working Paper 8243. Cambridge, MA: National Bureau of Economic Research.

Bard, Yonathan and Charles H. Sauer. 1981. "IBM Contributions to Computer Performance Modeling." *IBM Journal of Research and Development* 25:562–570.

Barzyk, Fred. 1999. "Updating the Hedonic Equations for the Price of Computers." Statistics Canada Prices Division Working Paper. Ottawa, ON: Statistics Canada.

Bell, C. Gordon. 1986. "RISC: Back to the Future?" *Datamation* 32(June):96–108.

Benkard, C. Lanier. 2001. *A Dynamic Analysis of the Market for Wide Bodied Commercial Aircraft.* Stanford, CA: Graduate School of Business, Stanford University.

Berndt, Ernst R. and Jack E. Triplett, eds. 1990. *Fifty Years of Economic Measurement.* Chicago, IL: University of Chicago Press.

Berndt, Ernst R., and Zvi Griliches. 1993. "Price Indexes for Microcomputers: An Exploratory Study." In Murray F. Foss, Marilyn Manser, and Allan H. Young, eds. *Price Measurements and Their Uses. Studies in Income and Wealth* 57:63–93. Chicago: University of Chicago Press for the National Bureau of Economic Research.

Berndt, Ernst R., Zvi Griliches, and Neal Rappaport. 1995. "Econometric Estimates of Prices in Indexes for Personal Computers in the 1990s." *Journal of Econometrics* 68(1995):243–268.

Berndt, Ernst R. and Neal J. Rappaport. 2001. "Price and Quality of Desktop and Mobile Personal Computers: A Quarter-Century Historical Overview." *American Economic Review* 91(2):268–273.

Berndt, Ernst R. and Neal J. Rappaport. 2002. "Hedonics for Personal Computers: A Reexamination of Selected Econometric Issues." Unpublished Paper.

Bhidé, Amar. 2006. "Venturesome Consumption, Innovation, and Globalization." Paper prepared for a joint conference of CESIFO and the Center on Capitalism and Society on "Perspectives on the Performance of the Continent's Economies." Venice. July 21-22.

Black, Sandra E. and Lisa M. Lynch. 2004. "What's Driving the New Economy?: The Benefits of Workplace Innovation." *Economic Journal.* 114(493):F97–F116.

Bleha, Thomas. 2005. "Down to the Wire," *Foreign Affairs* 84(3).

Blinder, Alan. 1997. "The Speed Limit: Fact and Fancy in the Growth Debate." *The American Prospect* 8(34).

Bloch, Erich and Dom Galager. 1978. "Component Progress: It's Effect on High-Speed Computer Architecture and Machine Organization." *Computer* 11(April):64–75.

Bosworth, Barry P. and Jack E. Triplett. 2001. "What's new about the New Economy? IT, Economic Growth, and Productivity." *International Productivity Monitor* 2(Spring):19–30.

Bosworth, Barry P. and Jack E. Triplett. 2003. *Services Productivity in the United States: Griliches' Services Volume Revisited.* Washington, D.C.: The Brookings Institution.

Bourot, Laurent. 1997. "Indice de Prix des Micro-ordinateurs et des Imprimantes: Bilan d'une rénovation." Working Paper of the Institut National De La Statistique Et Des Etudes Economiques (INSEE). Paris, France.

Brainard, Lael and Robert E. Litan. 2004. *"Off-shoring" Service Jobs: Bane or Boon and What to Do?* Brookings Institution Policy Brief 132. Washington, D.C.: The Brookings Institution.

Bresnahan, Timothy, Eric Brynjolfsson, and Lorin M. Hitt. 2002. "Information Technology, Workplace Organization, and the Demand for Skilled Labor: Firm-level Evidence." *Quarterly Journal of Economics* 117(1):339–376.

Bresnahan, Timothy and Manuel Trajtenberg. 1995. "General Purpose Technologies: Engines of Growth?" *Journal of Econometrics*, 65(1):83-108.

Bromley, D. Alan. 1972. *Physics in Perspective*. National Academy of Sciences. Washington, D.C.: National Academy Press.

Brooks, Frederick. 1975. *The Mythical Man Month: Essays on Software Engineering*. Reading, MA: Addison-Wesley Publishing Company.

Brynjolfsson, Eric and Lorin M. Hitt. 2000. "Beyond Computation: Information Technology, Organizational Transformation, and Business Performance." *Journal of Economic Perspectives* 14(4):23–48.

Brynjolfsson, Eric and Lorin M. Hitt. 2002. Computing Productivity: Firm-level Evidence." *Review of Economics and Statistics* 85(4):793–808.

Brynjolfsson, Eric and Brian Kahin, eds. 2000. *Understanding the Digital Economy*. Cambridge, MA: The MIT Press.

Bureau of Economic Analysis. 2001. "A Guide to the NIPAs." In *National Income and Product Accounts of the United States, 1929-97*. Washington, D.C.: U.S. Government Printing Office. Accessible at <http://www.bea.doc.gov/bea/an/nipaguid.pdf>.

Businessweek. 1997. "Deconstructing the Computer Industry." August 25.

Businessweek. 2005. "The Rise of India." December 8.

Butler Group. 2001. "Is Clock Speed the Best Gauge for Processor Performance?" *Server World Magazine*. September 2001. Accessed at *<http://www.serverworldmagazine.com/opinionw/2001/09/06_clockspeed.shtml>* on February 7, 2003.

Carr, Nicholas. 2005. "The End of Corporate Computing." *MIT Sloan Management Review* 46(3):67–73.

Cartwright, David W., Gerald F. Donahoe, and Robert P. Parker. 1985. "Improved Deflation of Computer in the Gross National Product of the United States." Bureau of Economic Analysis Working Paper 4. Washington, D.C.: U.S. Department of Commerce.

Cecchetti, Stephen G. 2002. "The New Economy and the Challenge for Macroeconomic Policy." Paper prepared for the conference, *The New Economy: What's new about it?* Texas A&M University. April 19.

Chandler, Alfred D. Jr. 2000. "The Information Age in Historical Perspective." In *A Nation Transformed by Information: How Information Has Shaped the United States from Colonial Times to the Present*. Alfred D. Chandler and James W. Cortada, eds. New York: Oxford University Press.

Choi, Soon-Yong and Andrew B. Whinston 2000. *The Internet Economy: Technology and Practice*. Austin, TX: SmartEcon Publishing.

Chow, Gregory C. 1967. "Technological Change and the Demand for Computers." *American Economic Review* 57(December):1117–1130.

Christiansen, Clayton. 1997. *The Innovator's Dilemma: When New Technologies Cause Great Firms to Fail*. Boston, MA: Harvard Business School Press.

Chwelos, Paul. 2003. "Approaches to Performance Measurement in Hedonic Analysis: Price Indexes for Laptop Computers in the 1990s." *Economics of Innovation and New Technology* 12(3):199–224.

Cohen, Stephen S., and John Zysman. 1988. *Manufacturing Matters: The Myth of the Post-Industrial Economy*. New York: Basic Books.

Cohen, Wesley M. and John Walsh. 2002. "Public Research, Patents and Implications for Industrial R&D in the Drug, Biotechnology, Semiconductor and Computer Industries." In National Research Council. *Capitalizing on New Needs and New Opportunities: Government-Industry Partnerships in Biotechnology and Information Technologies*. Washington, D.C.: National Academy Press.

Cole, Rosanne, Y. C. Chen, Joan A. Barquin-Stolleman, Ellen Dulberger, Nurhan Helvacian, and James H. Hodge. 1986. "Quality-Adjusted Price Indexes for Computer Processors and Selected Peripheral Equipment." *Survey of Current Business* 66(1):41–50.

Colecchia, Alessandra and Schreyer, Paul. 2002. "ICT investment and economic growth in the 1990s: is the United States a unique case? A comparative study of nine OECD countries." *Review of Economic Dynamics* 5(2):408–442.

Congressional Budget Office. 2002. *The Role of Computer Technology in the Growth of Productivity*. Washington, D.C.: Congressional Budget Office.

Corrado, Carol A., John Haltiwanger, and Daniel Sichel, eds. 2005. *Measuring Capital in a New Economy*. Chicago, IL: University of Chicago Press.

Council on Competitiveness. 2004. *Innovate America, Thriving in a World of Challenge and Change*, Washington, D.C.: Council on Competitiveness.

Council of Economic Advisors. 2001. *Annual Report*. Washington, D.C.: U.S. Government Printing Office.

Council of Economic Advisors. 2002. *Annual Report*. Washington, D.C.: U.S. Government Printing Office.

Crandall, Robert C. and Kenneth Flamm, eds. 1989. *Changing the Rules: Technological Change, International Competition, and Regulation in Communications*. Washington, D.C.: The Brookings Institution.

Cunningham, Carl, Denis Fandel, Paul Landler, and Robert Wright. 2000. *Silicon Productivity Trends*. International SEMATECH Technology Transfer #00013875A-ENG.

Dalén, Jorgen. 1989. "Using Hedonic Regression for Computer Equipment in the Producer Price Index." R&D Report. Statistics Sweden. Research-Methods-Development. 1989:25.

David, Paul. 2000. "Understanding Digital Technology's Evolution and the Path of Measured Productivity Growth: Present and Future in the Mirror of the Past." In Brynjolfsson, Eric and Brian Kahin, eds. *Understanding the Digital Economy*. Cambridge, MA: The MIT Press.

David, Paul. 2001. "Productivity Growth Prospects and the New Economy in Historical Perspective." *Eib Papers* 6(1):41–62.

DeLong, Bradford and Lawrence H. Summers. 2001. "The 'New Economy': Background, Historical Perspective, Questions, and Speculations." *Federal Reserve Bank of Kansas City Economic Review* 86(4):29–59.

Diewert, Irwin W. and Denis A. Lawrence. 2000. "Progress in Measuring the Price and Quantity of Capital." In Lawrence J. Lau, ed. *Econometrics and the Cost of Capital*. Cambridge, MA: The MIT Press.

Doms, Mark. 2003. "Communications Equipment: What Has Happened to Prices?" Federal Reserve Bank of San Francisco Working Paper, 2003-15.

Doms, Mark. 2004. "The Boom and Bust in Information Technology Investment." *Federal Reserve Bank of San Francisco Economic Review*, 19–34.

Doms, Mark. 2005. "Communications Equipment: What Has Happened to Prices?" in Corrado, Carol A., John Haltiwanger, and Daniel Sichel, eds. *Measuring Capital in a New Economy*. Chicago, IL: University of Chicago Press.

Dulberger, Ellen R. 1989. "The Application of a Hedonic Model to a Quality Adjusted Price Index for Computer Processors." In Dale W. Jorgenson and Ralph Landau, eds. *Technology and Capital Formation*: Cambridge, MA: The MIT Press.

Dulberger, Ellen R. 1993. "Sources of Price Decline in Computer Processors: Selected Electronic Components." In Murray Foss, Marilyn Manser, and Allan Young, eds. *Price Measurements and Their Uses*. Chicago: University of Chicago Press for the National Bureau of Economic Research.

Easterly, William. 2001. *The Elusive Quest for Growth*. Cambridge MA: The MIT Press.

The Economist. 1997. "Assembling the New Economy." September 11.

The Economist. 2000. "To Boldly Go…" March 23.

The Economist. 2000. "A Thinker's Guide." March 30.

The Economist. 2000. "Productivity on Stilts." June 8.
The Economist. 2000. "Performing Miracles." June 15.
The Economist. 2000. "Solving the Paradox." September 21.
The Economist. 2000. "Elementary, My Dear Watson." September 21.
The Economist. 2000. "Waiting for the New Economy." October 12.
The Economist. 2001. "The Great Chip Glut." August 11.
The Economist. 2001. "Productivity Growth (cont'd?)." September 6.
The Economist. 2003. "The New Geography of the IT Industry." July 17.
The Economist. 2003. "Overproductive and Underemployed." August 11.
The Economist. 2003. "Survey of the New Economy." September 11.
The Economist. 2003. "Between a Rock and a Hard Place." October 9.
The Economist. 2003. "Untangling the Local Loop." October 9.
The Economist. 2003. "Relocating the Back Office." December 11.
The Economist. 2004. "Innovative India." April 1.
The Economist. 2004. "Survey: A World of Work." November 11.
The Economist. 2005. "Moore's Law at 40." March 23.
The Economist. 2005. "How the Internet Killed the Phone Business." September 15.
Electronic News. 1999. "Sematech Adds 4 International Members." June 21.
Electronic News. 2005. "Samsung Faces $300M DoJ Fine for Price Fixing." October 13.
Ericson, R. and A. Pakes. 1995. "Markov-Perfect Industry Dynamics: A Framework for Empirical Work." *Review of Economic Studies* 62:53–82.
European Semiconductor Industry Association. 2005. *The European Semiconductor Industry 2005 Competitiveness Report*. Brussels, Belgium: European Semiconductor Industry Association. Accessed at *<http://www.eeca.org/pdf/final_comp_report.pdf>*.
Evans, Richard. 2002. "INSEE's Adoption of Market Intelligence Data for its Hedonic Computer Manufacturing Price Index." Presented at the Symposium on Hedonics at Statistics Netherlands. October 25.
Executive Office of the President. 2003. *The National Strategy to Secure Cyberspace: Cyberspace Threats and Vulnerabilities*. Washington, D.C.: Executive Office of the President.
Feenstra, Robert C., Marshall B. Reinsdorf, and Michael Harper. 2005. "Terms of Trade Gains and U.S. Productivity Growth." Paper prepared for NBER-CRIW Conference. July 25.
Fershtman, C. and A. Pakes. 2000. "A Dynamic Game with Collusion and Price Wars." *RAND Journal of Economics*. 31(2):207–236.
Fisher, Franklin M., John J. McGowan, and Joen E. Greenwood. 1983. *Folded, Spindled, and Multiplied: Economic Analysis and U.S. v. IBM*. Cambridge, MA: The MIT Press.
Flamm, Kenneth. 1988. *Creating the Computer*. Washington, D.C.: The Brookings Institution.
Flamm, Kenneth. 1989. "Technological Advance and Costs: Computers vs. Communications." In *Changing the Rules: Technological Change, International Competition, and Regulation in Communications*. Robert C. Crandall and Kenneth Flamm, eds. Washington, D.C.: The Brookings Institution.
Flamm, Kenneth. 1993. "Measurement of DRAM Prices: Technology and Market Structure." In Murray F. Foss, Marilyn E. Manser, and Allan H. Young, eds. *Price Measurements and Their Uses*. Chicago: University of Chicago Press.
Flamm, Kenneth. 1996. *Mismanaged Trade? Strategic Policy and the Semiconductor Industry*. Washington, D.C.: The Brookings Institution.
Flamm, Kenneth. 1997. *More for Less: The Economic Impact of Semiconductors*. San Jose, CA: Semiconductor Industry Association.
Flamm, Kenneth. 2003. "Microelectronics Innovation: Understanding Moore's Law and Semiconductor Price Trends." *International Journal of Technology, Policy, and Management* 3(2).
Flamm, Kenneth. 2003. "The New Economy in Historical Perspective: Evolution of Digital Technology." In *New Economy Handbook*. Derek C. Jones, ed. Academic Press.

Flamm, Kenneth. 2004. "Moore's Law and the Economics of Semiconductor Price Trends." In National Research Council, *Productivity and Cyclicality in Semiconductors: Trends, Implications, and Questions.* Dale W. Jorgenson and Charles W. Wessner, eds. Washington, D.C.: The National Academies Press.

Flamm, Kenneth. 2005. "The Coming IT Slowdown: Technological Roots and Economic Implications." Working Paper, LBJ School of Public Policy, University of Texas.

Flamm, Kenneth. 2006. "Economics of Innovation in the Microprocessor Industry." Working Paper. University of Texas at Austin. February.

The Financial Times. 2005. "The Internet's Next Big Talking Point: Why VoIP Telephony is Quickly Coming of Age." September 9.

Forbes. 2003. "Grove Says U.S. Is Losing Its Edge in High-Tech Sector." October 10.

Frankel, Jeffrey and Peter Orsag, eds. 2002. *American Economic Policy in the 1990s*. Cambridge, MA: The MIT Press.

Fransman, M. 1992. *The Market and Beyond: Cooperation and Competition in Information Technology Development in the Japanese System*. Cambridge, UK: Cambridge University Press.

Friedman, Thomas. 2005. *The World is Flat*. New York, NY: Farrar Straus & Giroux.

Gandal, Neil. 1994. "Hedonic Price Indexes for Spreadsheets and an Empirical Test for Network Externalities." *RAND Journal of Economics* 25.

Gereffi, Gary and Vivek Wadhwa. 2005. "Framing the Engineering and Outsourcing Debate: Placing the United States on a Level Playing Field with China and India." Duke University School of Engineering. December 12.

Gordon, Robert J. 1989. "The Postwar Evolution of Computer Prices." In Dale W. Jorgenson and Ralph Landau, eds. *Technology and Capital Formation*. Cambridge, MA: The MIT Press.

Gordon, Robert J. 1999. "Has the 'New Economy' Rendered the Productivity Slowdown Obsolete?" Paper presented at the Federal Reserve Bank of Chicago, June 9.

Gordon, Robert J., 2000. "Does the 'New Economy' Measure up to the Great Inventions of the Past?" *Journal of Economic Perspectives* 14(4). Northwestern University Working Paper.

Gordon, Robert J. 2000. "Interpreting the 'One Big Wave' in U.S. Long-Term Productivity Growth." NBER Working Paper 7752.

Gordon, Robert J. 2004. "Innovation and Future Productivity Growth: Does Supply Create its own Demand?" In Peter Cornelius, ed. *The Global Competitiveness Report 2002-3*. New York: Oxford University Press.

Gosling, James, Bill Joy, and Guy Steele. 1996. *The Java (TM) Language Specification*. New York: Addison-Wesley.

Griffith, P. 1993. "Science and the Public Interest." *The Bridge*. Washington, D.C.: National Academy of Engineering. (Fall):16.

Griliches, Zvi. 1960. "Measuring Inputs in Agriculture: A Critical Survey." *Journal of Farm Economics* 40(5):1398–1427.

Griliches, Zvi. 1961. "Hedonic Price Indexes for Automobiles: An Econometric Analysis of Quality Change." In George Stigler, ed. *The Price Statistics of the Federal Government*. New York: Columbia University Press.

Griliches, Zvi. 1994. "Productivity, R&D, and the Data Constraint." *American Economic Review* 94(2):1–23.

Grimm, Bruce, Brent R. Moulton, and David B. Wasshausen. 2005. "Information-Processing Equipment and Software in the National Accounts." In *Measuring Capital in the New Economy*. Carol Corrado, John Haltiwanger, and Daniel Sichel, eds. Chicago, IL: University of Chicago Press. Pp. 363–402.

Grindley, P., D. C. Mowery, and B. Silverman. 1994. "SEMATECH and Collaborative Research: Lessons in the Design of a High-Technology Consortia." *Journal of Policy Analysis and Management* 13.

Grossman, Gene and Elhannan Helpman. 1993. *Innovation and Growth in the Global Economy*. Cambridge, MA: The MIT Press.

Gowrisankaran, G. 1998. "Issues and Prospects for Payment System Deregulation." Working Paper. University of Minnesota.

Haavind, Robert. 2006. "Chipmaking's Tough Economic Road Ahead." *Solid State Technology*. March.

Hagel, J. and A. G. Armstrong. 1997. *Net Gain*. Cambridge, MA: Harvard Business School Press.

Halstead, Maurice H. 1977. *Elements of Software Science*. New York: Elsevier North Holland.

Handler, Philip. 1970. *Biology and the Future of Man*. London, UK: Oxford University Press.

Harhoff, Dietmar and Dietmar Moch. 1997. "Price Indexes for PC Database Software and the Value of Code Compatibility." *Research Policy* 24(4-5):509–520.

Hazlett, Thomas W. and George Bittlingmayer. 2003. "The Political Economy of Cable 'Open Access.'" *Stanford Technology Law Review* 4.

Helpman, Elhanan. 1998. "General Purpose Technologies and Economic Growth: Introduction." In *General Purpose Technologies and Economic Growth*. Elhanan Helpman, ed. Cambridge, MA: The MIT Press. Pp. 1–13.

Hobijn, Bart. 2001. "Is Equipment Price Deflation a Statistical Artifact?" *Federal Reserve Bank of New York Staff Report #139*. November.

Holdway, Michael. 2001. "Quality-Adjusting Computer Prices in the Producer Price Index: An Overview." Working Paper. Washington, D.C.: Bureau of Labor Statistics.

Holdway, Michael. 2002 "Confronting the Challenge of Estimating Constant Quality Price Indexes for Telecommunications Equipment in the Producer Price Index." Working Paper. Washington, D.C.: Bureau of Economic Analysis.

Hornstein, Andreas and Per Krusell. 2000. "The IT Revolution: Is It Evident in the Productivity Numbers?" *Federal Reserve Bank of Richmond Economic Quarterly* 86(4).

Horrigan, John Brendan. 2005. "Broadband Adoption at Home in the United States: Growing but Slowing." Washington, D.C.: Pew Internet and American Life Project.

Howell, Thomas. 2003. "Competing Programs: Government Support for Microelectronics." In National Research Council. *Securing the Future: Regional and National Programs to Support the Semiconductor Industry*. Charles W. Wessner, ed. Washington, D.C.: The National Academies Press.

Information Technology Association of America. 2004. "The Impact of Offshore IT Software and Services Outsourcing on the U.S. Economy and the IT Industry." Prepared by Global Insight.

International SEMATECH. 2004. "International Technology Roadmap for Semiconductors. Austin, TX. December. Available at <*http://public.itrs.net/*>.

International Telecommunication Union 2005. ITU Strategy and Policy Unit Newslog. August 8. Accessed at <*http://www.itu.int/osg/spu/newslog/CategoryView,category,Broadband.aspx*>.

Isenberg, David. 1997. "Rise of the Stupid Network." *Computer Telephony* (August):16–26.

Ishida, Haruhisa. 1972. "On the Origin of the Gibson Mix." *Journal of the Information Processing Society of Japan* 13(May):333–334 (in Japanese).

Jorgenson, Dale W. 2001. *Economic Growth in the Information Age*. Cambridge, MA: The MIT Press.

Jorgenson, Dale W. 2001. "Information Technology and the U.S. Economy." *American Economic Review* 91(1).

Jorgenson, Dale W. 2002. *Economic Growth in the Information Age-Volume 3*. Cambridge, MA and London, UK: The MIT Press.

Jorgenson, Dale W. 2002. "The Promise of Growth in the Information Age." The Conference Board Annual Essay.

Jorgenson, Dale W. 2004. "Information Technology and the World Economy." Leon Kozminsky Academy Distinguished Lecture. May 14.

Jorgenson, Dale W., Mun S. Ho, and Kevin J. Stiroh. 2002. "Projecting Productivity Growth: Lessons from the U.S. Growth Resurgence." *Federal Reserve Bank of Atlanta Economic Review* 87(3):1–13.

Jorgenson, Dale W., Mun S. Ho, and Kevin J. Stiroh. 2004. "Will the U.S. Productivity Resurgence Continue?" *Federal Reserve Bank of New York Current Issues in Economics and Finance* 10(13):1–7.

Jorgenson, Dale W., Mun S. Ho, and Kevin J. Stiroh. 2005. "Growth of U.S. Industries and Investments in Information Technology and Higher Education." In Carol A. Corrado, John Haltiwanger, and Daniel Sichel, eds. *Measuring Capital in a New Economy.* Chicago, IL: University of Chicago Press.

Jorgenson, Dale W., Mun S. Ho, and Kevin J. Stiroh. 2005. *Productivity, Volume 3: Information Technology and the American Growth Resurgence*, Cambridge, MA: The MIT Press.

Jorgenson, Dale W., J. Steven Landefeld, and William Nordhaus, eds. 2006. *A New Architecture for the U.S. National Accounts.* Chicago, IL: University of Chicago Press.

Jorgenson, Dale W. and Kevin J. Stiroh. 1999. "Productivity Growth: Current Recovery and Longer-term Trends." *American Economic Review,* 89(2):109–115.

Jorgenson, Dale W. and Kevin J. Stiroh. 2000. "Raising the Speed Limit: U.S. Productivity Growth in the Information Age." Brookings Papers on Economic Activity. Washington, D.C.: The Brookings Institution.

Jorgenson, Dale W. and Kevin J. Stiroh. 2002. "Raising the Speed Limit: U.S. Economic Growth in the Information Age." In National Research Council. *Measuring and Sustaining the New Economy.* Dale W. Jorgenson and Charles W. Wessner, eds. Washington, D.C.: National Academy Press.

Jorgenson, Dale W. and Eric Yip. 2000. "Whatever Happened to Productivity Growth?" In *New Developments in Productivity Analysis.* Charles R. Hulten, Edwin R. Rean, and Michael J. Harper, eds. Boston, MA: National Bureau of Economic Research.

Jovanovic, Boyan and Peter L. Rousseau. 2002. "Moore's Law and Learning-by-Doing." *Review of Economic Dynamics* 5:346–375.

Kessler, Michelle. 2002. "Computer Majors Down Amid Tech Bust." *USA Today*. October 8.

Knight, Kenneth E. 1966. "Changes in Computer Performance: A Historical View." *Datamation* (September):40–54.

Knight, Kenneth E. 1970. "Application of Technological Forecasting to the Computer Industry." In James R. Bright and Milton E.F. Schieman, *A Guide to Practical Technological Forecasting*. Englewood Cliffs, NJ: Prentice-Hall.

Knight, Kenneth E. 1985. "A Functional and Structural Measure of Technology." *Technological Forecasting and Technical Change* 27(May):107–127.

Koskimäki, Timo and Yrjö Vartia. 2001. "Beyond Matched Pairs and Griliches-Type Hedonic Methods for Controlling Quality Changes in CPI Sub-indices." Presented at Sixth Meeting of the International Working Group on Price Indices, sponsored by the Australian Bureau of Statistics, April.

Kozlow, Ralph and Maria Borga. 2004. "Offshoring and the US Balance of Payments." Washington, D.C.: Bureau of Economic Analysis.

Kuan, Jennifer. 2005. "Open Source Software as Lead User's Make or Buy Decision: A Study of Open and Closed Source Quality." Palo Alto, CA: Stanford Institute for Economic Policy Research.

Landefeld, J. Steven and Bruce Grimm. 2000. "A Note on the Impact of Hedonics and Computers on Real GDP." *Survey of Current Businesses* 80(12):17–22.

Landefeld, J. Steven and Barbara M. Fraumeni. 2001. "Measuring the New Economy." *Survey of Current Business* 81(3):23–40.

Landefeld, J. Steven and Robert P. Parker. 1997. "BEA's Chain Indexes, Time Series, and Measures of Long-Term Growth." *Survey of Current Business* 77(5):58–68.

Levine, Jordan. 2002. "U.S. Producer Price Index for Pre-Packaged Software." Presented at the 17th Voorburg Group Meeting. Nantes, France. September.

Levy, David and Steve Welzer. 1985. "An Unintended Consequence of Antitrust Policy: The Effect of the IBM Suit on Pricing Policy." Unpublished Paper. Rutgers University Department of Economics.

Levy, David L. 2005. "The New Global Political Economy." *Journal of Management Studies* 42(3):685.

Lim, Poh Ping and Richard McKenzie. 2002. "Hedonic Price Analysis for Personal Computers in Australia: An Alternative Approach to Quality Adjustments in the Australian Price Indexes." Paper presented at ZEW conference. Mannheim, Germany. April.

Litan, Robert E. and Roger G. Noll. 2003. "The Uncertain Future of the Telecommunications Industry." Brookings Working Paper. Washington D.C.: The Brookings Institution.

Litan, Robert E. and Alice M. Rivlin. 2000. "The Economy and the Internet: What Lies Ahead?" Washington, D.C.: Internet Policy Institute. Accessed at *<http://www.intenetpolicy.org/briefing/litan_rivlin.html>*.

LoPiccolo, Phil. 2006. "The Six Billion Dollar Gap." *Solid State Technology*. February.

Macher, Jeffrey T., David C. Mowery, and David A. Hodges. 1999. "Semiconductors." In National Research Council. *U.S. Industry in 2000: Studies in Competitive Performance*. David C. Mowery, ed. Washington, D.C.: National Academy Press.

Maney, Kevin. 2003. "Music Industry Doesn't Know What Else to Do as It Lashes Out at File-Sharing." *USA Today*. September 9.

Mann, Catherine. 2003. "Globalization of IT Services and White Collar Jobs: The Next Wave of Productivity Growth." *International Economics Policy Briefs* PB03-11. December.

Mann, Catherine. 2004. "The U.S. Current Account, New Economy Services, and Implications for Sustainability." *Review of International Economics* 12(2):262–276.

Mann, Catherine L. Forthcoming. *High Technology and the Globalization of America*.

Marciniak, John J., ed. 1994. *Encyclopedia of Software Engineering*. New York, NY: John Wiley & Sons.

Martin, Brookes and Zaki Wahhaj. 2000. "The Shocking Economic Impact of B2B." *Global Economic Paper, 37*. Goldman Sachs. February 3.

Maxwell, Kim. 1998. *Residential Broadband: An Insider's Guide to the Battle for the Last Mile*. Hoboken, NJ: John Wiley & Sons.

McKinsey Global Institute. 2001. *U.S. Productivity Growth 1995-2000: Understanding the Contribution of Information Technology Relative to Other Factors*. Washington, D.C.: McKinsey & Company.

McKinsey Global Institute. 2003. *Off-shoring: Is it a Win-Win Game?* San Francisco, CA: McKinsey & Company.

McKinsey Global Institute. 2005. *The Emerging Global Labor Market*. Washington D.C.: McKinsey & Company.

Michaels, Robert. 1979. "Hedonic Prices and the Structure of the Digital Computer Industry." *The Journal of Industrial Economics* 27 (March):263–275.

Moch, Dietmar. 2001. "Price Indices for Information and Communication Technology Industries: An Application to the German PC Market." Center for European Economic Research (ZEW) Discussion Paper No. 01-20. Mannheim, Germany: ZEW.

Moore, Gordon E. 1965. "Cramming More Components onto Integrated Circuits." *Electronics* 38(8) April.

Moore, Gordon E. 1975. "Progress in Digital Integrated Circuits." *Proceedings of the 1975 International Electron Devices Meeting* 11–13.

Moore, Gordon E. 1997. "The Continuing Silicon Technology Evolution Inside the PC Platform." *Intel Developer Update*. Issue 2.

Moore, Gordon E. 2003. "No Exponential if Forever . . . but We Can Delay Forever." Santa Clara, CA: Intel Corporation.

Moylan, Carol. 2001. "Estimation of Software in the U.S. National Income and Product Accounts: New Developments." Paris, France: Organisation for Economic Co-operation and Development.

National Academy of Sciences, National Academy of Engineering, Institute of Medicine. 1993. *Science, Technology and the Federal Government. National Goals for a New Era.* Washington, D.C.: National Academy Press.

National Academy of Sciences, National Academy of Engineering, Institute of Medicine. 2007 Forthcoming. *Rising Above the Gathering Storm: Energizing and Employing America for a Brighter Economic Future.* Washington, D.C.: The National Academies Press.

National Advisory Committee on Semiconductors. 1992. *A National Strategy for Semiconductors: An Agenda for the President, the Congress, and the Industry.* Washington, D.C.: National Advisory Committee on Semiconductors.

National Institute of Standards and Technology. 2002. "The Economic Impacts of Inadequate Infrastructure for Software Testing." Planning Report 02-3. Gaithersburg, MD: U.S. Department of Commerce.

National Research Council. 1995. *Standards, Conformity Assessment, and Trade into the 21st Century.* Washington, D.C.: National Academy Press.

National Research Council. 1996. *Conflict and Cooperation in National Competition for High-Technology Industry.* Washington, D.C.: National Academy Press.

National Research Council. 1999. *The Advanced Technology Program: Challenges and Opportunities.* Charles W. Wessner, ed. Washington, D.C.: National Academy Press.

National Research Council. 1999. *Industry-Laboratory Partnerships: A Review of the Sandia Science and Technology Park Initiative.* Charles W. Wessner, ed. Washington, D.C.: National Academy Press.

National Research Council. 1999. *New Vistas in Transatlantic Science and Technology Cooperation.* Charles W. Wessner, ed. Washington, D.C.: National Academy Press.

National Research Council. 1999. *The Small Business Innovation Research Program: Challenges and Opportunities.* Charles W. Wessner, ed. Washington, D.C.: National Academy Press.

National Research Council. 2001. *The Advanced Technology Program: Assessing Outcomes.* Charles W. Wessner, ed. Washington, D.C.: National Academy Press.

National Research Council. 2001. *Building a Workforce for the Information Economy.* Washington, D.C.: National Academy Press.

National Research Council. 2001. *Capitalizing on New Needs and New Opportunities: Government-Industry Partnerships in Biotechnology and Information Technologies.* Charles W. Wessner, ed. Washington, D.C.: National Academy Press.

National Research Council. 2001. *A Review of the New Initiatives at the NASA Ames Research Center.* Charles W. Wessner, ed. Washington, D.C.: National Academy Press.

National Research Council. 2001. *Trends in Federal Support of Research and Graduate Education.* Stephen A. Merrill, ed. Washington, D.C.: National Academy Press.

National Research Council. 2002. *At What Price? Conceptualizing and Measuring Cost-of-Living and Price Indexes.* Washington D.C.: National Academies Press.

National Research Council. 2002. *Measuring and Sustaining the New Economy.* Dale W. Jorgenson and Charles W. Wessner, eds. Washington, D.C.: National Academy Press.

National Research Council. 2002. *Partnerships for Solid-State Lighting.* Charles W. Wessner, ed. Washington, D.C.: National Academy Press.

National Research Council. 2003. *Government-Industry Partnerships for the Development of New Technologies: Summary Report.* Charles W. Wessner, ed. Washington, D.C.: The National Academies Press.

National Research Council. 2003. *Securing the Future: Regional and National Programs to Support the Semiconductor Industry.* Charles W. Wessner, ed. Washington, D.C.: The National Academies Press.

National Research Council. 2004. *Productivity and Cyclicality in Semiconductors: Trends, Implications, and Questions.* Dale W. Jorgenson and Charles W. Wessner, eds., Washington, D.C.: The National Academies Press.

National Research Council. 2004. *The Small Business Innovation Research Program: Program Diversity and Assessment Challenges*. Charles W. Wessner, ed. Washington, D.C.: The National Academies Press.

National Research Council, 2005. *Deconstructing the Computer*. Dale W. Jorgenson and Charles W. Wessner, eds. Washington, D.C.: The National Academies Press.

National Research Council. 2005. *Getting Up to Speed: The Future of Supercomputing*. Washington, D.C.: The National Academies Press.

National Research Council. 2005. *Globalization of Materials R&D: Time for a National Strategy.* Washington, D.C.: National Academies Press.

National Research Council. 2005. *Policy Implications of International Graduate Students and Postdoctoral Scholars in the United States*. Washington, D.C.: The National Academies Press.

National Research Council. 2006. *Software, Growth, and the Future of the U.S. Economy.* Dale W. Jorgenson and Charles W. Wessner, eds. Washington, D.C.: The National Academies Press.

National Research Council. 2006. *The Telecommunications Challenge: Changing Technologies and Enduring Policies.* Charles W. Wessner, ed. Washington, D.C.: The National Academies Press.

Nelson, Richard, ed. 1993. *National Innovation Systems*. New York: Oxford University Press.

Nelson, R. A., T. L. Tanguay, and C. C. Patterson. 1994. "A Quality-adjusted Price Index for Personal Computers." *Journal of Business and Economics Statistics* 12(1):23–31.

The New York Times. 2003. "Good Economy. Bad Job Market. Huh?" September 14.

The New York Times. 2004. "Financial Firms Hasten Their Move to Outsourcing." August 14.

Nikkei Microdevices. 2001. "From Stagnation to Growth: The Push to Strengthen Design." January.

Nordhaus, William D. 2002. "Productivity Growth and the New Economy." *Brookings Papers on Economic Activity* 2:211–44.

Nordhaus, William D. 2002. "The Progress of Computing." New Haven, CT: Yale University. March 4.

Nuechterlein, Jonathan E. and Philip J. Weiser. 2005. *Digital Crossroads : American Telecommunications Policy in the Internet Age.* Cambridge, MA: The MIT Press.

Okamoto, Masato and Tomohiko Sato. 2001. "Comparison of Hedonic Method and Matched Models Method Using Scanner Data: The Case of PCs, TVs and Digital Cameras." Presented at Sixth Meeting of the International Working Group on Price Indices, sponsored by the Australian Bureau of Statistics. April.

Oliner, Stephen. 2000. The Resurgence of Growth in the Late 1990s: Is Information Technology the Story?" *Journal of Economic Perspectives* 14(4):3–22.

Oliner, Stephen. 2002. "Information Technology and Productivity: Where Are We Now and Where Are We Going?" *Federal Bank of Atlanta Economic Review* 87(3):15–44.

O'Mahoney, Mary and Bart van Ark. 2003. *EU Productivity and Competitiveness: An Industry Perspective: Can Europe Resume the Catching-up Process?* Luxembourg: Office for Official Publications of the European Communities.

O'Reilly, Tim. 2005. "What Is Web 2.0—Design Patterns and Business Models for the Next Generation of Software." Accessed at <*http://www.oreillynet.com/pub/a/oreilly/tim/news/2005/09/30/what-is-web-20.html*> on September 30, 2005.

Organisation for Economic Co-operation and Development. 2000. *Is There a New Economy? A First Report on the OECD Growth Project.* Paris, France: Organisation for Economic Co-operation and Development.

Organisation for Economic Co-operation and Development. 2002. *Information Technology Outlook 2002—The Software Sector.* Paris, France: Organisation for Economic Co-operation and Development. P. 105.

Organisation for Economic Co-operation and Development. 2003. *ICT and Economic Growth.* Paris, France: Organisation for Economic Co-operation and Development.

Organisation for Economic Co-operation and Development. 2003. Report of the OECD Task Force on Software Measurement in the National Accounts. Statistics Working Paper 2003/1. Paris, France: Organisation for Economic Co-operation and Development.

PC Magazine. 2006. "AMD to Build Factory in New York." June 26.

PC World Magazine. 2003. "20 Years of Hardware." March.

Paganetto, Luigi, ed. 2004. *Knowledge Economy, Information Technologies, and Growth*. Burlington, VT: Ashgate.

Pakes, G. E. 1966. *Physics Survey and Outlook*. National Academy of Sciences. Washington D.C.: National Academy Press.

Pakes, Ariel. 2001. "A Reconsideration of Hedonic Price Indices with an Application to PCs." Harvard University. November.

Parker, Robert P. and Bruce Grimm. 2000. "Recognition of Business and Government Expenditures for Software as Investment: Methodology and Quantitative Impacts, 1959-98." Washington, D.C.: Bureau of Economic Analysis.

Porter, Michael. 2004. "Building the Microeconomic Foundations of Prosperity: Findings from the Business Competitiveness Index." In X. Sala-i-Martin, ed. *The Global Competitiveness Report 2003-2004*. New York: Oxford University Press.

Poulsen, Kevin. 2004. "Software Bug Contributed to Blackout." *Security Focus*. February 11.

President's Council of Advisors on Science and Technology. 2004. *Sustaining the Nation's Innovation Ecosystems*. Washington, D.C.

Pritchard, Stephen. 2003. "Munich Makes the Move." *Financial Times*. October 15.

Prud'homme, Marc and Kam Yu. 2002. "A Price Index for Computer Software Using Scanner Data." Unpublished working paper, Prices Division, Statistics Canada. Ottawa, ON.

Raduchel, William. 2006. "The Economics of Software." In National Research Council, *Software, Growth, and the Future of the U.S. Economy*. Dale W. Jorgenson and Charles W. Wessner, eds. Washington, D.C.: The National Academies Press.

Rao, H. Raghaw and Brian D. Lynch. 1993. "Hedonic Price Analysis of Workstation Attributes." *Communications of the Association for Computing Machinery (ACM)* 36(12):94–103.

Robertson, Jack. 1998. "Die Shrinks Now Causing Logic Chip Glut." *Semiconductor Business News*. October 15.

Ruttan, Vernon. 2001. *Technology, Growth, and Development*. New York: Oxford University Press.

Samuelson, Paul. 2004. "Where Ricardo and Mill Rebut and Confirm Arguments of Mainstream Economists Supporting Globalization." *Journal of Economic Perspectives* 18(3).

Schaller, Robert R. 1999. "Technology Roadmaps: Implications for Innovation, Strategy, and Policy." Ph.D. Dissertation Proposal, Institute for Public Policy, George Mason University.

Schaller, Robert R. 2002. "Moore's Law: Past, Present, and Future." Accessed at <*http://www.njtu.edu.cn/depart/xydzxx/ec/spectrum/moore/mlaw.html*> on July 2002.

Schlender, Brent. 1999. "The Edison of the Internet." *Fortune*. February 15.

SEMI. 2005. "Semiconductor Equipment and Materials: Funding the Future." Accessed at <*http://content.semi.org/cms/groups/public/documents/homepervasive/p036611.pdf*>. Press Release. October.

Semiconductor Industry Association. 2004. *International Technology Roadmap for Semiconductors: Update*. Accessed at <*http://www.itrs.net/Common/2004Update/2004Update.htm*>.

Semiconductor Industry Association. 2005. *International Technology Roadmap for Semiconductors*. Accessed at <*http://www.itrs.net/Common/2005ITRS/Home2005.htm*>.

Semiconductor Industry Association. 2005. "Semiconductor Industry Association Says U.S. Could Lose Race for Nanotechnology Leadership." Accessed at <*http://www.sia-online.org/pre_release.cfm?ID=355*>. Press Release. March 16.

Semiconductors International. 2000. "Sematech Forms International Sematech." March.

Shapiro, Carl and Hal R. Varian. 1999. *Information Rules*. Boston, MA: Harvard Business School Press.

Sharpe, William F. 1969. *The Economics of the Computer*. New York, NY and London, UK: Columbia University Press.

Sichel, Daniel E. 1997. *Computer Revolution: An Economic Perspective*. Washington, D.C.: The Brookings Institution.

Siebert, Horst. 2002. *Economic Policy Issues of the New Economy*. Heidelberg, Germany and New York: Springer.

Sigurdson, Jon. 1986. *Industry and State Partnership in Japan: The Very Large Scale Integrated Circuits (VLSI) Project*. Lund, Sweden: Research Policy Institute.

Sirgudson, Jon. 2004. "VSLI Revisited—Revival in Japan." Working Paper No. 191. Tokyo, Japan: Institute of Innovation Research of Hitotsubashi University.

Soete, Luc. 2001. "The New Economy: A European Perspective." In Daniele Archibugi and Bengt-Ake Lundvall, eds. *The Globalizing Learning Economy*. Oxford, UK and New York: Oxford University Press.

Solow, Robert M. 1987. "We'd Better Watch Out." *New York Times Book Review*. July 12.

Solow, Robert M., Michael Dertouzos, and Richard Lester. 1989. *Made in America*. Cambridge, MA: The MIT Press.

Spencer, W. J. and P. Grindley. 1993. "SEMATECH After Five Years: High Technology Consortia and U.S. Competitiveness." *California Management Review* 35.

Spencer, W. J. and T. E. Seidel. 2004. "National Technology Roadmaps: The US Semiconductor Experience." In National Research Council. *Productivity and Cyclicality in Semiconductors: Trends, Implications, and Questions*. Charles W. Wessner, ed. Washington, D.C.: The National Academies Press.

Spencer, William, Linda Wilson, and Robert Doering. 2005. "The Semiconductor Technology Roadmap." *Future Fab International* 18. January 12.

Statistics Finland. 2000. "Measuring the Price Development of Personal Computers in the Consumer Price Index." Paper for the Meeting of the International Hedonic Price Indexes Project. Paris, France. September 27.

Stiroh, Kevin J. 2001. "Investing in Information Technology: Productivity Payoffs for U.S. Industries. *Federal Reserve Bank of New York Current Issues in Economics and Finance* 7(6):1–6.

Stiroh, Kevin J. 2002. "Are ICT Spillovers Driving the New Economy?" *Review of Income and Wealth* 48(1).

Stiroh, Kevin J. 2002 "Information Technology and the U.S. Productivity Revival: What Do the Industry Data Say?" *American Economic Review* 92(5):1559–1576.

Stiroh, Kevin J. 2002. "Measuring Information Technology and Productivity in the New Economy." *World Economics* 3(1):43–58.

Stoneman, Paul. 1976. *Technological Diffusion and the Computer Revolution: The U.K. Experience*. Cambridge, UK: Cambridge University Press.

Telecommunications Industry Association. 2002. "TIA Tells FCC That Unbundling Rules Discourage Broadband Investment." Press Release. April 8.

Temple, Jonathan. 2002. "The Assessment: The New Economy." *Oxford Review of Economic Policy* 18(3):241–264.

Triplett, Jack E. 1985. "Measuring Technological Change with Characteristics-Space Techniques." *Technological Forecasting and Social Change* 27:283–307.

Triplett, Jack E. 1986. "The Economic Interpretation of Hedonic Methods." *Survey of Current Business* 66(1):36-40. January.

Triplett, Jack E. 1989. "Price and Technological Change in a Capital Good: A Survey of Research on Computers." In Dale W. Jorgenson and Ralph Landau, eds. *Technology and Capital Formation*. Cambridge, MA: The MIT Press.

Triplett, Jack E. 1996. "High-Tech Productivity and Hedonic Price Indexes." in Organisation for Economic Co-operation and Development. *Industry Productivity*. Paris, France: Organisation for Economic Co-operation and Development.

Triplett, Jack E. 1997. "The Solow Productivity Paradox: What Do Computers Do to Productivity?" Paper presented at the Conference on Service Sector Productivity and the Productivity Paradox, Ottawa, ON. April 11–12.

Triplettt, Jack E. 1999. "Did the U.S. Have a New Economy?" Paper presented for the Association de Compatible Nationale 9th Conference on National Accounting.

Triplett, Jack E. 1999. "Economic Statistics, the New Economy, and the Productivity Slowdown." *Business Economics*. January.

Triplett, Jack E. 1999. "The Solow Productivity Paradox: What Do Computers Do to Productivity?" *Canadian Journal of Economics* 32(2):309–334.

Triplett, Jack E. 2004. *Handbook on Hedonic Indexes and Quality Adjustments in Price Indexes: Special Application to Information Technology Products*. Paris: Organisation for Economic Co-operation and Development.

Triplett, Jack E. and Barry Bosworth. 2002. "Baumol's Disease Has Been Cured: IT and Multifactor Productivity in U.S. Service Industries." Presented at the Brookings Workshop "Services Industry Productivity: New Estimates and New Problems." March 14. Accessible at *<http://www.brook.edu/dybdocroot/es/research/projects/productivity/workshops/20020517.htm>*.

Tyson, Laura. 1992. *Who's Bashing Whom? Trade Conflict in High Technology Industries*. Washington, D.C.: Institute for International Economics.

U.S. Display Consortium. 2003. "The Global FPD Industry—2003: An In-depth Overview and Roadmap." San Jose, CA.

United States Congress. 2001. "Information Technology and the New Economy." Washington, D.C.: Joint Economic Committee Study.

van Ark, Bart, Robert Inklaar, and Robert H. McGuckin. 2002. "Changing Gear, Productivity, ICT and Service Industries: Europe and the United States." Brookings Seminar Paper on Productivity in Services. Washington, D.C.: The Brookings Institution.

van Mulligen, Peter Hein. 2002. "Alternative Price Indices for Computers in the Netherlands Using Scanner Data." Prepared for the 27th General Conference of the International Association for Research in Income and Wealth, Djurhamn, Sweden.

van Welsum, Desirée and Xavier Reif. 2005. "Potential Off-shoring: Evidence from Selected OECD Countries." Paris, France: Organisation for Economic Co-operation and Development.

Vatter, Harold G. and John F. Walker. 2001. "Did the 1990s Inaugurate a New Economy?" *Challenge!* 44(1):90–116.

The Wall Street Journal. 2003. "Searching for Computing's Kilowatt." July 17.

The Wall Street Journal. 2004. "Outsourcing May Create U.S. Jobs." March 30.

The Wall Street Journal. 2005. "Lopsided Immigration Policy Could Induce Brain Drain." June 22. P. A17.

The Wall Street Journal. 2006. "'Net Neutrality' Snags Overhaul of Telecom Laws." June 29. P. A7.

The Washington Post. 2004. "Election Campaign Hit More Sour Notes." February 22.

Whelan, Karl. 2002. "Computers, Obsolescence, and Productivity." *Review of Economics and Statistics* 84(4):445–462.

Wolff, Alan Wm., Thomas R. Howell, Brent L. Bartlett, and R. Michael Gadbaw, eds. 1992. *Conflict Among Nations: Trade Policies in the 1990s*. San Francisco, CA: Westview Press.

World Semiconductor Trade Statistics. 2000. *Annual Consumption Survey*.

Woroch, Glenn. 2002. "Local Network Competition." In *Handbook of Telecommunications Economics*. Martin Cave, Sumit Majumdar, and Ingo Vogelsang, eds. New York, NY: Elsevier.

Wyckoff, Andrew W. 1995. "The Impact of Computer Prices on International Comparisons of Labour Productivity." *Economics of Innovation and New Technology* 3:277–293.